Fruit and Vegetable Biotechnology

Fruit and Vegetable Biotechnology

– Author –
Bibhuti Dutta Jha

Wordpen Academics
Publishing Science That Matter

ISBN: 978-93-90438-52-5 (Hardback)

Published by : **WORDPEN ACADEMICS**
4736/23, Ansari Road, Darya Ganj,
New Delhi – 110 002
Phone: +91-011-35004336
E-mail: info@wordpenacademics.com

Printed at **:** **Replika Press Pvt. Ltd.**

Exclusive Distributor: ***Bio-Green Books, New Delhi***

Preface

Biotechnology can be seen as an imprecise term since the harnessing of any biological process could justifiably be called biotechnology. In food processing it could reasonably be applied to processes as long established as bread making and brewing. However, the revolution in our understanding of the molecular mechanisms underlying the processes of life, in particular our understanding of DNA, has resulted in the potential to manipulate those mechanisms for our requirements. Transfer and expression of foreign genes in plant cells, now routine practice in several laboratories around the world, has become a major tool to carry out gene expression studies and to obtain plant varieties of potential agricultural interest. The capacity to introduce and express diverse foreign genes in plants, first described for tobacco in 1984, has been extended to many species. Transgenic crops such as tomato, cotton, maize, soybean, etc., are now available for human consumption and by complementing traditional methods of crop improvement (and thus becoming an integral part of agriculture), they will have a profound impact on food production, economic development and on the development of a sustainable agricultural system during the 21st century.

Genetic information is carried in the linear sequence of nucleotides in DNA. Its expression involves the translation of the linear sequence of specific regions of DNA existing in the nucleus of the cell (called coding regions or genes) into a colinear sequence of amino acids (proteins). As an intermediate step, however, DNA must be copied into a different type of polynucleotide known as ribonucleic acid (RNA) which retains all the information of the DNA sequence from which it was copied. Single-stranded RNA molecules are synthesized by a process known as DNA transcription which is regulated by interactions between DNA sequences located upstream of the gene (promoters) and proteins (transcription factors). Thousands of RNA transcripts can be made from the same DNA segment in a given cell. Many of these RNA molecules undergo major chemical changes before they leave the nucleus to serve as the messenger RNA (mRNA) molecules that direct the synthesis of proteins in the cytosol. Fragments of DNA can be amplified by a process called DNA cloning which consists in inserting the DNA into a plasmid or a bacterial virus and then growing these in bacterial (or yeast) cells. Plasmids are small circular molecules of DNA that occur naturally in bacteria, where they replicate as independent units. As these bacteria divide, the plasmid also replicates to produce an enormous number of copies of the cloned DNA fragment. Although restricted genomic DNA fragments can be cloned to produce genomic libraries, cDNA libraries are most frequently used to isolate and characterize genes necessary for the production of genetically engineered plants. cDNA libraries represent the information encoded in the mRNA of a particular tissue or organism. mRNA molecules are exceptionally labile and difficult to amplify in their natural form. For this reason, the information encoded by the mRNA is converted into a stable DNA duplex (cDNA) via enzymatic reactions catalysed by reverse transcriptase and DNA polymerase I, and then is inserted into a self-replicating plasmid. The resulting heterogeneous population of cDNA molecules collectively encodes virtually all of the mRNAs sysnthesized by the cell. Once the information is available in the form of a cDNA library, individual processed segments of the original genetic information can be isolated and examined with relative ease. Development of procedures in cell biology to regenerate plants from single cells and the discovery of techniques to transfer and express foreign genes to plant cells provided the prerequisite for the practical use of genetic engineering in crop improvement. The essential requirements in a gene transfer system for production of transgenic plants are the availability of a target tissue having cells competent for both plant regeneration and transformation, a method to introduce DNA into cells, a procedure to select transformed cells and a system to regenerate plants from the transformed cells at a satisfactory rate.

The book begins with an analysis of the methods available to the biotechnologist. It considers the range of traits that have been the subject of modification. The book discusses the modification of such agronomic traits as fruit quality and resistance to various kinds of environmental stress, as well as the use of molecular markers in plant breeding.

– Bibhuti Dutta Jha

Contents

1

Agricultural Biotechnology

Definition of Biotechnology for the Purposes of the Forum

Firstly, how is biotechnology defined for the purposes of the Forum? According to the Convention on Biological Diversity (CBD), biotechnology means "any technological application that uses biological systems, living organisms, or derivatives thereof, to make or modify products or processes for specific use". Interpreted in this broad sense, the definition covers many of the tools and techniques that are commonplace today in agriculture and food production.

Interpreted in a narrow sense, as is often done and as is done in the Forum, biotechnology mainly covers technological applications involving reproductive biology or, secondly, the manipulation, or use, of the genetic material of living organisms for specific uses. This definition covers a wide range of diverse technologies including, for example, the use of molecular DNA markers, gene manipulation and gene transfer, vegetative reproduction (crops and forest trees), embryo transfer and freezing (livestock) and triploidization (fish).

Background to the Establishment of the Forum at FAO

FAO was founded in 1945 with a mandate to raise levels of nutrition and standards of living, to improve agricultural productivity, and to better the condition of rural populations. It is an intergovernmental organization with 183 member countries. One of the most important tasks that FAO carries out is to collect, analyse, interpret and disseminate relevant information. FAO serves as a clearing-house, providing farmers, scientists, government planners, traders and non-governmental organizations (NGOs) with the information they need to make rational decisions on planning, investment, marketing, research and training. FAO should play an active part in disseminating information and promoting information exchange regarding biotechnology. It is important that member countries know which biotechnologies are available, what they can be used for, how and in which wider strategy they can be applied, and what the cost-benefit implications of using them are.

The global population size has passed the six billion mark and is increasing by roughly 80 million annually. Almost all population growth is in developing countries. Since the amount of agricultural land available is limited, the increases in food production needed to feed the world's growing population must come from increasing the amount of food produced per hectare. Biotechnology, which is a collection of diverse tools that can be applied to many areas of food and agriculture, may play a role here.

This collection includes scientific tools (such as genetic modification) that are sometimes considered to be highly controversial. The tools may pose ethical problems and require substantial debate among policy-makers, researchers and the public at large. Particularly in some areas of agricultural biotechnology (e.g. involving cultivation of genetically modified crops), the debate has become quite polarized and there is therefore an increasing need for quality, balanced, neutral and factual information.

To consider specifically the background to the establishment of the Forum, the biennial meeting of FAO's Committee on Agriculture (COAG), held in Rome from 25-29 January 1999, was of key importance because, among other areas, it set the direction for FAO's future involvement in biotechnology. The report of the Committee "stressed FAO's role in providing a forum for countries to monitor food and agriculture biotechnologies".

At its 116th session (14-19 June 1999), the FAO Council subsequently endorsed the COAG report, stating that it "appreciated the need for FAO to have a coherent programme on agricultural biotechnology to assist Member Nations in obtaining the full benefits of new developments while minimizing risks. FAO's role as a forum for the discussion of issues and for standard-setting, and as an 'honest broker' of quality science-based information, through mechanisms such as the International Plant Protection Commission (IPPC) and Codex Alimentarius, was underscored in general, and in relation to biotechnology in particular".

Later, at the 30th session of the FAO Conference (12-23 November 1999), Members stated that one of the substantive areas to which they attached particular importance was the active contribution of FAO to current debates on biotechnology and genetically modified organisms. It was therefore in this spirit that FAO established the Forum. It is coordinated by the FAO Inter-Departmental Working Group on Biotechnology (IDWGB) that was established in 1999 following the recommendations of the 1999 COAG meeting.

Operation and Structure of the Forum

The Forum has an open structure that allows various parties-policy-makers, people from universities, NGOs, the public, etc.-to discuss and exchange views and experiences about specific issues concerning biotechnology and its applications in the animal, fishery, forestry and plant sectors in developing countries. The principal activity of the Forum is to run moderated e-mail conferences (each lasting roughly two months) about specific topics concerning biotechnology in food and agriculture for developing countries. To register for any conference, individuals must first be members of the Forum.

The topics all have biotechnology as the core subject and may cover themes such as biosafety, public/private agricultural research, biodiversity, capacity-building, food safety, poverty alleviation, benefit sharing, intellectual property rights and food production. The emphasis is on developing countries. As the Forum covers the broad range of activities found within the area of food and agriculture, it covers topics both of specific relevance to those interested in the animal, crop, fish or forestry sectors or of general relevance to all sectors.

The Forum was officially launched on 9 March 2000. The launching was marked by sending an e-mail "letter of invitation" to a list of people and institutions that might have been interested in this initiative. An important source of e-mail addresses was a report on biotechnology networks in developing countries, prepared for FAO's Research and Technology Development Service (SDRR) in September 1999. The list also included Permanent Representatives to FAO and all FAO country representatives and was supplemented by additional addresses provided by members of the IDWGB. In the letter of invitation, people were requested to also forward the information to anyone that they considered might be interested.

The number of Forum members rose to over 700 within the first month, to over 1 000 after three months and to nearly 1 300 by the time the sixth conference was finished more than one year later. Once they joined, very few people left the Forum. Forum members may not send messages to each other (although if they register for a conference they may send a message to all other participants in the conference) and may only receive messages from the Forum Administrator, who is responsible for all contact with the Forum members. They have so far (November 2001) received 31 messages i.e. roughly two a month from the Forum Administrator.

Forum members are not automatically registered for any e-mail conference, but instead have to do this themselves. Thus, they participated to varying degrees in the different conferences. Some did not register for any conference, but instead only received key documents from the Forum Administrator. Others instead, registered for several conferences and received all the e-mail messages posted.

Table 1: Sequence of key events regarding the Forum, including the number of Forum members at each date

Date	*Event*	*No. Forum Members*
9 March 2000	1) Forum launched 2) Forum website launched	0
20 March	Conference 1 begins	519
25 April	Conference 2 begins	814
26 May	Conference 1 ends	932
12 June	Conference 3 begins	1 008
30 June	Conference 2 ends	1 086
1 August	Conference 4 begins	1 158
25 August	Conference 3 ends	1 182
8 October	Conference 4 ends	1 205
1 November	Conference 5 begins	1 217
17 December	Conference 5 ends	1 208*
20 March 2001	Conference 6 begins	1 240
14 May	Conference 6 ends	1 282

* The drop in numbers is due to removing some non-valid e-mail addresses in December

When the Forum was launched, a website to complement and support the Forum was also launched. The website was implemented in collaboration with FAO's information management group, WAICENT. Note, however, that the primary communication medium of the Forum is e-mail, so to be a Forum member and participate actively in any of its conferences, the only thing that is required is an e-mail account. The website merely gathers together in one place all the information about the Forum, as well as all the documents and individual messages related to the different conferences. The website has been recognized as a valuable resource. It was selected by the Internet Scout Project for inclusion in the Scout Report (26 May 2000), a weekly current awareness publication that highlights new internet resources of interest to researchers and educators; was chosen as a "Hot Pick" in the Netwatch section of the journal Science; as well as the "site of the Day" by New Scientist (29 December 2000).

It was a conscious decision to operate the Forum with e-mail as the base communication medium (rather than, for example, running the conferences on the web) to try and facilitate participation from developing countries. Although both typically require access to a computer, modem, phone line and an account with an internet service provider, full internet access with browsing on the web tends to be more expensive and more difficult in practice than simply receiving and sending e-mail messages. The analyses carried out, showing that individuals from developing countries were actively involved in the Forum conferences but very seldom visited the Forum website, strongly support this decision.

Individuals wishing to join the Forum, have to register themselves. This is done by sending an e-mail message to an automatic FAO mail server. Using the server, people can automatically subscribe or unsubscribe themselves from the Forum, or they can receive messages previously posted by the Forum Administrator. Registration is also possible from the Forum website.

Operation of the Individual e-mail Conferences

The six conferences were operated in the same way.

Before a Conference

Before a given e-mail conference began, a Background Document, two to five pages in length, was prepared. In this publication, the six Background Documents are included. As the conferences took place in a time span covering over one year, the documents were written at different stages from March 2000 to March 2001.

The aim of the Background Document is to give an easily-understandable description of the conference theme, enabling potential participants to have a basic grasp of some of the main aspects of the theme. For example, the Background Document to Conference 2, on the appropriateness of current biotechnologies for the forestry sector in developing countries, provided a brief summary of the kinds of biotechnologies currently available for the forestry sector; some key elements or current trends in the forestry sector in developing countries and finally, certain factors that should be considered in the discussions. Before a conference began, the Background Document was sent to Forum members. In the same e-mail message, they were invited to join the conference and given instructions about doing so. They were requested to carefully read the document if wishing to participate in the conference.

During a Conference

Involvement of Forum members in each conference is governed by the "Rules of the Forum" and "Guidelines for Participation in E-mail Conferences" that Forum members receive on joining the Forum. These specify, *inter alia,* that

- they should introduce themselves briefly in their first posting to a conference;
- they should exercise tolerance and respect toward other participants whose views may differ from their own, and remain courteous at all times;
- they should not submit messages longer than 600 words;
- people represent only themselves i.e. that "regardless of whether they identify the entity by whom they are employed, participants are assumed to be speaking in their personal capacity unless they explicitly state that their contribution

represents the views of their organization. For this reason, participants should not quote the postings of other participants as representing the views of the organizations to which those other participants belong".

Each conference was moderated by the Forum Administrator. The Moderator's role is to screen all messages before they are posted to ensure that they follow the rules and guidelines of the Forum and that they are relevant to the theme of the conference. In addition, the Moderator plays an active role in the conference by ensuring that messages are understandable and, where appropriate, providing additional information of benefit to participants.

Roughly 95 percent of messages received by the Moderator during FAO working hours were posted to the conference within an hour of receipt. Those received after working hours were usually posted first thing the following morning. Only a small minority of messages was refused for posting, coming mainly in Conference 1. Messages were refused primarily because they were not directly relevant to the theme of the conference. When required, IDWGB members provided technical support to the Moderator.

Midway through each conference, a brief Update Document was written and sent to Forum members, summarizing the kinds of messages posted, the subjects dealt with and pointing out some areas that should be addressed in the remaining time available. In some cases, as in Conference 1, more than one Update was written.

After a Conference

After a conference is finished, two Summary Documents are written. The first version is longer (five to eleven pages), more detailed and contains references to specific e-mail messages. The second version is shorter (one to two pages) and does not contain references. Both documents attempt to provide an easily-readable summary of the main arguments and concerns discussed during the conference, based on the messages posted by the participants. In this publication, the longer versions of the Summary Documents are provided. References are made to specific e-mail messages that can be viewed on the Forum website.

The Six Conferences

The first conference began less than two weeks after the Forum was launched. It was the first of a four-conference block on the theme of the appropriateness of currently available biotechnologies in the crop, forestry, animal and fishery sectors respectively for food and agriculture in developing countries. The themes of the fifth and sixth conferences were chosen based on the interest shown in them by participants during the early conferences. The fifth conference dealt with the implications of agricultural biotechnology for hunger and food security in developing countries, while the sixth examined the impact of intellectual property rights on food and agriculture in developing countries. The titles and start/end dates of the six conferences are as follows:

Conference 1 (20 March to 26 May 2000): How appropriate are currently available biotechnologies in the crop sector for food production and agriculture in developing countries?

Conference 2 (25 April to 30 June 2000): How appropriate are currently available biotechnologies for the forestry sector in developing countries?

Conference 3 (12 June to 25 August 2000): The appropriateness, significance and application of biotechnology options in the animal agriculture of developing countries.

Conference 4 (1 August to 8 October 2000): How appropriate are currently available biotechnologies for the fishery sector in developing countries?

Conference 5 (1 November to 17 December 2000): Can agricultural biotechnology help to reduce hunger and increase food security in developing countries?

Conference 6 (20 March to 14 May 2001): The impact of intellectual property rights (IPR) on food and agriculture in developing countries.

The conferences thus dealt with specific separate themes, although always remaining within the general area of biotechnology in food and agriculture in developing countries. The conferences attracted different audiences, discussed different topics (although they often overlapped) and each had different characteristics. Some general figures from the conferences are provided in Table 2.

Table 2: Number of people that registered for each conference, number of messages posted and the number of weeks the conference lasted

Conference	*Theme*	*No. members registered*	*No. messages posted*	*Duration of conference (wks)*
1	Crops	306	138	10
2	Forestry	167	32	9.5
3	Livestock	235	42	11
4	Fishery	149	26	10
5	Hunger/Food security	258	118	6.5
6	IPR	265	50	8

Limitations of the conferences

Language

The six conferences took place in English only. Thus the Background and Summary Documents for each conference, as well as all messages from the Moderator and by participants (with the exception of a couple of messages transmitted in both English and French in the livestock sector conference) were in English. This affects the kind of audience and participants in the different conferences, making it difficult for individuals lacking the English language to contribute to the conferences and to make their opinions/experiences known. For example, in the fisheries sector conference, there were no messages from some of the developing countries that have active programmes in fisheries biotechnology (Brazil, China and Cuba) and language might explain their absence. Nevertheless, messages posted in the Forum conferences came from nearly 50 different countries throughout the world, many of which do not have English as their main language.

Electronic Communication

When people attend a "traditional" conference, the list of participants typically includes those who are invited and those who either pay for themselves or who are paid to attend by their employer. There is often a restriction on the maximum number of attendees. There is thus a certain process of selection where (language considerations apart) a number of people interested in a particular subject may not be present to provide their input. However, with any type of conference, the quality of the discussions and outputs depends on the participants. The aim of the Forum is to allow a wide range of parties to discuss and exchange views and experiences about specific issues concerning biotechnology. The medium for communication is e-mail and, as with a traditional conference, this also involves problems of selectivity.

Even though the Forum is free and open to everyone it requires, however, in the most typical case, electricity, a phone connection and a computer with a modem. There are thus large differences between and within countries regarding access to these new communication technologies.

The UNDP Human Development Report 2001 showed there is a large "digital divide" in the world today. Figures from the report indicated that, in the year 2000, nearly seven percent of the world population was using internet but that 79 percent of internet users lived in Organisation for Economic Cooperation and Development (OECD) countries. Furthermore, the percentage of the population with internet use ranged from 28 percent in high-income OECD countries to 0.4 percent in sub-Saharan Africa or South Asia. In addition, the report also provided some information on the "digital divide" *within* countries, pointing out that internet users are mainly:

- urban and located in certain regions (e.g. in China, only four million of 600 million people in 15 poorly connected provinces are internet users while two major cities, Shanghai and Beijing, have five million);
- better educated and wealthier (e.g. in Chile, 89 percent of users have third-level education);

- young (e.g. in China, 84 percent are under 35);
- male (e.g. 86 and 62 percent of users are male in Ethiopia and Latin America, respectively).

These are important limitations that should be kept in mind when reading the summaries from the conferences.

The Benefits of Agricultural Biotechnology

Agricultural biotechnology is any technique in which living organisms, or parts of organisms are altered to make or modify agricultural products, to improve crops, or develop microbes for specific uses in agricultural processes. Simply put, when the tools of biotechnology are applied to agriculture, it is termed as "agricultural biotechnology". Genetic engineering is also a part of agricultural biotechnology in today's world. It is now possible to carry out genetic manipulation and transformation on almost all plant species, including all the world's major crops.

Plant transformation is one of the tools involved in agricultural biotechnology, in which genes are inserted into the genetic structure or genome of plants. The two most common methods of plant transformation are Agrobacterium Transformation-methods that use the naturally occurring bacterium; and Biolistic Transformation - involving the use of mechanical means. Using any of these methods the preferred gene is inserted into a plant genome and traditional breeding method followed to transfer the new trait into different varieties of crops.

Production of food crops has become much cheaper and convenient with the introduction of agricultural biotechnology. Specific herbicide tolerant crops have been engineered which makes weed control manageable and more efficient. Pest control has also become more reliable and effective, eliminating the need for synthetic pesticides as crops resistant to certain diseases and insect pests have also been engineered. Phytoremediation is the process in which plants detoxify pollutants in the soil, or absorb and accumulate polluting substances out of the soil. Several crops have now been genetically engineered for this purpose for safe harvest and disposal, and improvement of soil quality.

According to the USDA (United States Department of Agriculture)'s National Agricultural Statistics Service (NASS), in reference to a section specific to the major biotechnology derived field crops, out of the whole crop plantings in the United States in 2004, biotechnology plantings accounted for about 46 percent for corn, 76 percent for cotton, and 85 percent for soybeans.

Modern agricultural biotechnology has now become a very well-developed science. The use of synthetic pesticides that may be harmful to man, and pollute groundwater and the environment, has been significantly lessened with the introduction of genetically engineered insect-resistant cotton. Herbicide-tolerant soybeans and corn have also enabled the use of reduced-risk herbicides that break down more quickly in soil. These are nontoxic to plants or animals, and herbicide-tolerant crops help preserve topsoil from erosion since they thrive better in no-till or reduced tillage agriculture systems. Papayas resistant to the ringspot virus were also developed through genetic engineering, which saved the U.S. papaya industry.

Agricultural biotechnology may also be helpful in improving and enhancing the nutritious quality of certain crops. For example, enhancing the levels of beta-carotene in canola, soybean, and corn improves oil compositions, and reduces vitamin A deficiencies in rice. There are also researches going on in the field of biotechnology to produce crops that will not be affected by harsh climates or environments and that will require less water, fertilizer, labour etc. This would greatly reduce the demands and pressures on land and wildlife.

Biotechnology in Agriculture, Forestry and Fisheries

In this document, biotechnology is defined as "any technique that uses living organisms to make or modify a product, to improve plants or animals, or to develop microorganisms for specific uses" (after the Office of Technology Assessment, US Congress). The document focuses on the development and application of modern biotechnology which is based on the use of new enabling techniques of recombinant-DNA technology, often referred to as "genetic engineering".

Biotechnology has great potential to influence and benefit agriculture, forestry and fisheries. Modern techniques of biotechnology offer the potential of moving any cloned gene from any organism into any other organism and confer much greater precision and speed in achieving results as compared to conventional techniques. In conjunction with conventional technologies, modern biotechnology holds promise of increased and sustained productivity, efficient processing for improved product diversification and utilization, adaptation of product quality to functional requirements, and decreased reliance on agrochemicals and other external inputs. It may also promote better conservation and use of genetic resources, and environmentally friendly management of natural resources. However, the number of marketable products and their influence at the farm level still seems to be limited, but is likely to increase in the next decade.

Biotechnology also poses certain challenges. These are largely determined by how, where and when it finds application. In general, the fast-paced research, predominantly funded by private-sector investment, and use of intellectual property rights in industrialized countries are seen as evidence that the application of biotechnology will hold the key to competitiveness and comparative advantage in many fields, including agriculture and food.

Biotechnology, with its vast potential and challenges, is thus of the utmost importance to agricultural development, and hence to FAO. The science of developing biotechnological tools cannot make a distinction between developing and developed countries. However, the application of such tools in the development process requires preconditions which are easily found in many developed countries, but which hardly exist in most developing countries. The current paths of research and development have given rise to concern that the disparity in harnessing biotechnologies for agricultural and economic development may increase between industrialized and developing countries.

The preamble to the FAO Constitution requires that the Organization work towards raising levels of nutrition and the standards of living of all peoples, securing improvements in the efficiency of the production and distribution of all food and agricultural products, improving the conditions of rural populations and thus contributing towards an expanding world economy. Therefore, one of the major tasks for the Organization is to ensure that the benefits of biotechnology will be shared by people in the north and the south, in both large and small, and rich and poor countries.

In recent years, various FAO bodies and conferences such as the Tenth Session of FAO's Committee on Agriculture, 1991; the FAO Conference 1989; Sessions of the FAO Commission on Plant Genetic Resources, 1989 and 1991; FAO/CTA International Symposium on Plant Biotechnologies for Developing Countries, 1989; and the 20th FAO Regional Conference for Asia and the Pacific, 1990, have strongly recommended that FAO, as the lead UN specialized agency for food and agriculture, must respond to the potential and challenges of modern biotechnology for agriculture, forestry and fisheries.

This paper, therefore, examines the pattern of biotechnological development in developed and developing countries, the application, impacts and potential of biotechnologies in agriculture, forestry and fisheries. It then analyses some of the issues and concerns in biotechnology such as trade substitution, biosafety, and intellectual property rights, with special reference to developing countries. Finally, it elaborates the elements of FAO's policy and strategies on biotechnology.

Developments in Biotechnology

There exists a gradient of biotechnologies, depending on the degree of sophistication, complexity, stage of development and application. The lower side of the gradient comprises simpler but widely used techniques such as in vitro culture, rhizobium technology, fermentation. The upper side of the gradient includes advanced techniques involving genetic engineering. The gradient of biotechnologies may be matched with the gradient of national capabilities, economic investment and efforts to provide the possibility of choosing the appropriate techniques and approaches with the most positive impacts. The pattern of use and development of biotechnologies in developed and developing countries thus varies considerably.

The Pattern in Developed Countries

In the industrialized countries, a new pattern of biotechnological research funding has emerged. With the availability of property right protection of biotechnologies and prospects of vast markets for biotechnology products and the techniques, the bulk of the research is funded, carried out, and controlled by the private sector. For instance, since 1976 when GENENTECH, the first biotechnology company, was established, the new techniques have spawned a variety of industries that now comprise more than 400 start-up firms, more than 200 established firms that have diversified into biotechnology, and more than 200 supply firms in the United States alone. The new biotechnology industry in the United States produced pharmaceutical, diagnostic tests and agricultural products worth close to US$2 billion in 1990. A similar trend is seen in Europe and Japan. It is estimated that about 60 percent of the biotechnology research and development funding in industrialized countries is from the private sector. Thus, the private sector has been a major force in enhancing the capability of these countries in this field.

Research institutions in the public sector are now generally required to raise a substantial part of their budget from non-government sources, via contractual research, licensing agreements and royalties. This is tending to increase secrecy over research findings and to hinder free scientific communication. University professors, researchers and government institution scientists are increasingly becoming entrepreneurial and are entering private industry.

Another important trend is that large multinational corporations are purchasing smaller seed and biotechnology companies and diversifying their holdings. This allows them to develop a package sale of chemicals, seeds and equipment. Heavy involvement of the private sector and market considerations greatly influence the topics and commodities chosen for research. Major crops, commodities and farming systems of great socio-economic importance to the developing world, but of little international market importance, do not figure in the biotechnology research agenda of industrialized countries. Furthermore, these countries are keen to reduce their production costs, increase the productivity, quality and value of their products and, thus, improve their overall competitiveness in the world market.

The Patterns in Developing Countries

Biotechnology facilities are being established in most developing countries. However, the level of research, development and use of biotechnology for agriculture, forestry and fisheries in the developing countries is generally far below the level in the industrialized countries. Among developing countries, the status varies considerably.

A few, such as Brazil, Mexico, India, China and The Republic of Korea, have sought to gain full scientific and technological capacity, especially in agricultural biotechnology. Others, such as Indonesia, Malaysia, the Philippines and Thailand and a few countries of Latin America, have built the capacity to apply biotechniques and develop biotechnologies useful for their agriculture and food industries. The participation of the private sector in gaining biotechnological capacity is not significant in most of these countries.

Many developing countries have inadequate funding, poor human resources and limited access to information, resulting in a relatively low-level capacity for research and technology development and exploitation, especially in the field of modern biotechnology research, which is costly and requires highly trained personnel. Most developing countries are vague as to their immediate aims in agricultural biotechnology. Few have appropriate proprietary-right protection systems or mechanisms to increase their access to protected techniques and products. Furthermore, there is negligible involvement of the private sector, which accentuates the problem of insufficient attention to biotechnology.

One of the major constraints on biotechnological development in developing countries is the poor quality and extent of higher education in frontier sciences, especially molecular biology. Besides, there are no or very weak links between universities and research institutions to reinforce each other and to utilize synergistically the scarce trained personnel resources.

The university experts play a negligible role in national policy formulation, including that on biotechnology. Furthermore the universities are generally not oriented for efficiency in commercialization and are therefore not able to market biotechnology products of their own or others' inventions.

Some international and donor-country public sector institutions strive to ensure that attention is given to solving, through biotechnology, the food and agriculture problems of developing countries. Among this group are the International Agricultural Research Centres (IARCs) located or operating in developing countries. Capacity exists in several of these centres for utilization of advances in biotechnologies to solve certain problems of plant production and protection and of livestock production and health, and capacity is being built in others.

For example, the International Rice Research Institute (IRRI) is incorporating modern molecular biology in its rice research activities. Its programme is linked closely to the Rockefeller Foundation Biotechnology Network. The World Bank, the

International Service for National Agricultural Research (ISNAR) and the Australian Government undertook a major study on the likely impact of modern biotechnology on agriculture. This study is being followed up by the participants by an expansion in the World Bank lending for biotechnology and by ISNAR's establishing an Intermediary Biotechnology Service to provide advisory services to national agricultural research systems.

Applications, Impact and Potential

Biotechnological research and development are moving at a very fast rate. For instance, three years ago, transgenics in rice, especially in japonica rices, were considered rather difficult because of the problem of regeneration through protoplast culture. But today, a large number of transgenics both in indica and japonica types are available and are being tested at various stages. It is therefore difficult to predict the potential and impact of a given technology beyond five years or so. This section gives a brief account of the current level of technology and refers to the developments likely to occur in the next three to five years.

The potential of genetic engineering in medicine has received much attention. The potential for advances in agriculture, forestry and fishery are similarly bright. Biotechniques are already being used to create new strains of crop plants, new plant and animal diagnostic products, animal vaccines, biological pesticides and herbicides, other biological control agents, and modifications in domestic animals used for food production. In some cases, applications are being held in check by the need for still more research to ensure that there are no harmful effects (or, if there are, they are greatly outweighed by the benefits) and by the slow pace of evolution of regulations governing the release of genetically modified organisms or products for general use or acquisition.

Biotechnological research, especially genetic engineering, on problems of field and tree crops, forest species and fisheries has a relatively short history. The knowledge of biological aspects of species and ecosystems needs to be expanded in may cases before new biotechnologies can be applied effectively. Even where transfer of one or more genes can be achieved reliably, expression of the transferred gene(s) may not occur in the expected way. The search for and identification and cloning of useful genes, effective vector systems, methods of gene transfer and promoter mechanisms should be intensified.

A perspective on both the difficulties and intensity of the research effort suggests that diagnostic tools will speed solutions in the next five to ten years to certain disease problems in several crop, forest, animal and fish species, such as black sigatoka of banana and plantain, and virus/viroid diseases of coconut and rice. Improved tolerance to certain physical stresses, such as heat tolerance in potatoes, may also be achieved, but prospects for improved resistance of varieties and clones to most abiotic stresses are longer term.

Crops

Modern biotechnologies can add greater precision and speed to plant breeding. Transgenics have already been reported in 40 crop plants, including maize, rice, soybean, cotton, rapeseed, potato, sugar beet, tomato, potato and alfalfa, but the new varieties are yet to be used commercially. Near-future opportunities for commercial exploitation include vegetables and fruits (potato, tomato, cucumber, cantaloupe and squash), followed by legumes (alfalfa) and oilseed crops (oilseed rape). A good number of the transgenics are herbicide-resistant plants whose widespread use is somewhat controversial.

Wide use is currently made of tissue culture techniques for micropropagation of elite clones and for freeing planting materials of pathogens. Monoclonal antibodies are also in use as diagnostic aids in the detection and identification of viruses and viroids.

Another culture and microspore culture giving rise to haploids are being used in variety improvement to facilitate and accelerate breeding. Molecular maps and markers are being widely used to identify genes of interest to accelerate conventional breeding programmes. Efficient biological nitrogen fixation systems and strains for efficient utilization of soil nutrients are being genetically engineered. Other long-term objectives are the genetic manipulations of photosynthesis pattern and the production of hybrid seed through apomixis. A very distant possibility is providing nitrogen fixation capacity to cereals.

Livestock

Among agricultural and allied fields, animal production and health have benefited the most from biotechnology, although practical use of transgenic livestock is only a future possibility. Wide use of monoclonal antibodies for efficient diagnostics, leading to safe and specific treatments of animal diseases, is a major breakthrough.

Through genetic engineering, vaccines for the prevention of viral, bacterial, and parasitic animal diseases are rendered more effective and safe. Tailored vaccines exist for pig scours, chicken bursar disease and cattle tickborne diseases. Pathogen-specific vaccines are attractive goals. Other interesting possibilities are endocrine-directed vaccines to stimulate twinning in beef cattle, immunocastration, livestock growth rate increase and vaccines that compensate for various stress-induced production losses.

Advances in genetic engineering will also facilitate the production of male-only populations of screwworms, tsetse flies, ticks and various other ectoparasites for use in the sterile release technique of control or eradication. Furthermore mammalian tissue culture may replace whole animals in the 1990s for toxicity testing of certain chemicals. The culture technique can also be exploited for studying and analyzing pesticide metabolism and for herbicide pre-screening. In vitro fertilization and embryo sexing techniques have considerably increased the use of embryo transfer techniques for cattle breeding and trade. The value of the approach will be further enhanced if embryo cloning techniques can be reliably employed. Microbial and enzymatic treatment of roughages and genetic engineering of rumen bacteria both have great potential to improve animal nutrition. Growth hormones can be produced by genetically engineered microorganisms in quantities and at the low cost necessary for widespread use to speed and increase milk and lean meat production. Biotechnological tools (embryo culture, gene cloning, etc.) may also be used for conservation of genetic resources.

Forestry

Successful regeneration through micropropagation or somatic embryogenesis has been achieved for about 100 forest species although, for most, considerable development work would be required before commercial propagation could be contemplated. In breeding programmes incorporating clonal testing, inclusion of a micromultiplication phase may facilitate more rapid deployment of superior genotypes than that afforded by sequential multiplication by cuttings. This will particularly be the case when gene mappling techniques are sufficiently advanced to permit accurate identification of superior genotypes without replicated clonal testing in the field, especially within a breeding line where substantial genetic disequilibrium exists. Furthermore, gene mapping may allow eventual identification of valuable genes, including those that contribute to quantitative traits.

As regards the application of genetic engineering, many major traits of commercial importance to forestry are under polygenic control, and much remains to be learned about the operation of the genes involved before this technique can have a major impact for these characters. An earlier application of genetic engineering in some forestry operations may be the introduction of genes, some already available, known to confer insect of disease resistance. Other applications of biotechnology with obvious value for forestry, but as yet not widely supported by experimental successes, include: in vitro manipulation of the maturation state, e.g. the promotion of early flowering to reduce generation intervals, in vitro selection for traits such as disease resistance and tolerance to salinity; and the use of haploid cultures.

Fisheries

Two broad areas of biotechnology exist within the fisheries sector, namely natural products biotechnology (including mostly marine species); and aquaculture biotechnology. Commercially valuable products such as pharmaceutical compounds, pigments, oils, sterols, alginates and agarose are being extracted from micro-and macro-algae in many parts of the world. Aquatic invertebrates are currently being screened for biologically active compounds that may have, inter alia anti-carcinogenic or antiviral properties, whereas marine bacteria are currently utilized in treatments of oil spills and holding tanks in tanker ships.

Within aquaculture, induced increases in the chromosome complement (polyploidy) of commercially important species such as salmon and oysters have increased the growth and marketability of these species. Genetically engineered microbes can be used to produce fish growth hormones, which might then be used to improve feed conversion and growth rate. Synthetic reproductive hormones are produced commercially and used to regulate fecundity, breeding cycles, growth rates and sex determination in certain cultured species. In attempts to increase desirable culture qualities such as growth rate, disease resistance, temperature tolerance and marketability, transgenic fishes containing introduced genes from other species have been produced on an experimental scale. The application of advanced biotechnology to the fisheries sector is a relatively recent practice. Therefore, continued research can be expected to reveal additional commercially viable applications and products within this sector.

Processing and Product Quality and Uses

Biotechnology could have great potential for improving the quality and diversifying the uses of agricultural, forestry and fisheries products. This technology could be used for processing products by both traditional and new methods in order to:

- improve the value, safety and quality of processed and unprocessed foods, such as genetically engineered tomatoes which soften more slowly and which could be processed at lower temperatures, resulting in a higher proportion of solids to liquid in canned tomatoes;
- enhance non-food uses of farm products and thereby increase their market value;
- develop new uses for farm products, especially those often producing surpluses, including energy production from biomass and recycling of wastes and by products. An example of such an application is found in ethanol production from cassava and sugar cane;
- produce algal cells from, for instance, Spirulina or Chlorella in mass culture for proteins;
- produce enzymes on a commercial scale, especially using the microbial system, as substitutes for enzymes isolated from animal or plant sources. For example,

chymosin, the enzyme used in the manufacture of cheese, traditionally taken from the cow, can now be made from specially engineered microorganisms. Mass production of amino acids from molasses and other agricultural resources is another example. These amino acids can improve the nutritional value of food and animal feed;

- produce and use new probes for rapid detection and accurate identification of specific microbial pathogens in food, especially diagnostics for staphylococcal enterotoxins, and others; and
- enhance productivity through the application of newly developed processes such as the immobilized enzyme and plant tissue culture systems which are being used to produce useful secondary metabolites. Examples are diosgenin, a female contraceptive commonly used in China and India and shikonin, a natural pigment used for lipstick in Japan.

Biotechnology is being used in a number of countries to characterize the indigenous germplasm variability for therapeutic and biologically important substances. The use of biotechnology for the removal of toxic and antinutritional factors and for improving the quality of food products is gaining momentum. These techniques include the elimination or reduction of cyanogenic compounds in cassava and the incorporation of high protein in potato and sweet potato.

2

Prospects and Limitations of Agricultural Biotechnologies

Introductory Remarks

It is now generally that the practice of plant breeding will move forward as progress is made in knowledge and technology. Breeders in the future will not only be forced to hone their traditional skills but will also have to integrate them with knowledge and experience to coincide with more recent advancements and technologies derived from cellular and molecular concepts and approaches.

For worldwide crops such as potato, rice, maize and in the future wheat, as well as for high value vegetable crops, the future of breeders depends upon the use of a combination of breeding practices (traditional breeding, genetic engineering and tissue culture). In addition to an increasing reliance on breeding, agriculture will also depend on biocontrol to complement and to make possible the use of chemicals that are compatible with intelligent management of the natural resources essential for a sustainable yet highly productive agriculture.

Thus most aspects of agriculturally relevant biotechnology are covered. Plant tissue culture techniques not only provide essential ways for the clonal propagation of many agriculturally important crops(e.g woody plants ornamentals and vegetables) but they are also the basis for the production of transgenic plants and are at the forefront of recent studies in plant physiology, developmental biology and biochemistry.

Spectacular progress in our knowledge of plant sciences, including those relevant to agriculture and breeding, can be expected from the integration of more traditional sciences with molecular and cellular concepts and techniques. A prime example is seen in the impact molecular genome analysis is already having on traditional breeding (via so called map-based breeding) but also on research and molecular breeding (e.g, via map based gene-cloning) This rosy description of the future of genetic engineering is richly tempered in this book by a realistic assessment of the many problems that still need to be faced before what is predicted can be accomplished. It was therefore wise to ask scientists well known for their "no-nonsence" approach such as Ingo Potrykus and his colleagues to write about genetic engineering of crop plants.

To quote from their introduction: " it has been possible to develop the state of the art of gene transfer to plants to a level at which many of the major crop plants are accessible to the technique and for which we have good reason to believe that there is no basic biological principle that will prevent gene transfer to a specific crop plant " Yet these authors clearly identify the considerable research and development that is still needed before one can, in many but not all instances, consider large scale applications.

A recurrent point made and well documented in various chapters of this book is that genetic engineering-because it is not limited by the natural barriers that ensure the maintenance of species and prevent transfer of genes between unrelated organisms-is an extremely powerful way to increase the genetic variability available to breeders. The present rise in interest to acquire new knowledge via fundamental research can best be illustrated by reminding ourselves of recent progress in our understanding of the molecular physiology of development, of the mechanism of action of old and newly discovered phytohotmones and of growth signals used by microorganisms that are pathogens or symbionts of plants (such as for instance lipo-chitooligosaccharides).

Biotechnology can contribute to making traditional agriculture more productive while at the same time more sustainable. It can also help achieve the goals of "non-food" and or cash crops agriculture as has been described in the various chapters dealing with crop improvement and metabolite production. Hopefully sufficient attention will be paid to regional crops of importance by local populations in developing countries.

Finally topics such as "Biofertilization" and "Bioremediation" are convincingly dealt with Topics such as "Biotechnology of Farm Animals" and "Marine Biotechnology" as well as the unavoidable but all important "Legal and Public Aspects" round out the all inclusive title of Agricultural Biotechnologies.

Some topics have already come to blossom and illustrate well the potential of present day plant sciences. There is the breakthrough in the cloning and characterization of resistance genes along with significantly improved understanding of the nature of avirulence genes and elicitors. Many hypothetical models predict that resistance genes should be involved in the signal transduction pathways linking pathogen derived signals with the regulated expression of defence genes and of programmed cell death (hypersensitivity reaction). The first available data on resistance genes show that this concept may well be correct in its very general sense, even if all resistance genes do not code for elicitor receptors.

We may safely predict that in the near future we will know more details of the molecular mechanisms underlying plant-pathogen resistance mechanisms. Several elicitor receptors will soon be isolated and characterized along with genes involved in localized pathogen-induced programmed cell death. Plant breeders will make use of isolated resistance genes to produce crops with improved resistance to a variety of pathogens and pests. The content of this book fully justifies the conclusion that Biotechnologies are bound to play an ever growing role in agriculture not only in the long term future, but also presently, and in the near future.

Fortunately the progress in applied plant biotechnology corresponds with and is in fact stimulating fundamental scientific progress. This is important because the real driving force behind this science is the success of applied plant biotechnology such as. 1) plant protection based on the expression of several different introduced genes coding for different principles acting on the same targets and thereby providing a possible solution to the problem of the emergence of resistant pathogens and pests, 2) new quality traits as well as future crops producing tailor-made non-food products (lipids, carbohydrates, and biodegradable thermoplastics).

Only if plant biotechnology becomes a commercial as well as an environmental success will sufficient support be available for this fascinating research.

Plants and Agriculture

Human population has grown steadily since the advent of agriculture, nearly 12,000 years ago, which ensured a continuing and reliable supply of food. The most rapid growth occurred during the past two centuries during which time agriculture became highly industrialized, public health improved and there were increases in incomes and international trade in food grains. Despite the age-old human concerns about balancing population and food supplies and many population control measures the worlds population is projected to continue to grow into the early decades of the 21st century, stabilizing at 9-11 billion sometimes during 2030-2050. Much of this growth will occur in the already overpopulated underdeveloped and poorer regions of Africa, Asia and Latin America, which will be home to nearly 90% of the human population.

Food shortages were common during the early and middle years of the 20th century. The introduction of Green Revolution varieties of wheat, rice and maize during the mid-1960s reversed this trend and helped avoid major food shortages by keeping increases in food productivity slightly ahead of population growth. During this period, both China and India, the two most populous nations with chronic food shortages, became bet food exporters. Realistically however such increases in food productivity can not be sustained indefinitely. It is not surprising, therefore that increases in food productivity have begun to decline during the past few years. Furthermore, improved economic conditions in China and India have created greater demand for better and varied food products, particularly poultry and meat, which require greater supplies of feed grains. As a result, both China and India have now become net importers of food. World food reserves have declined from a high of 77 days to less that 50 days.

With the current trends of population growth and agricultural production the demand for food in the most populous parts of the World will double by the year 2025, and nearly triple by 2050, Increases in food productivity of this magnitude can not be brought about in such a short period of time by conventional breeding, especially when some of out most important crops are approaching the physiological limits of productivity, not by increasing the amount of arable land, which accounts for 97% of all food production in the world..

Arable land, which is finite and comprises about 3% of the earth's surface, is deteriorating and decreasing as a result of soil erosion, salinization, over cultivation, and acidification. These factors, combined with expected increases in population, will actually decrease the global per capita arable land from the current 0.28 hectare to 0.17 hectare by the year 2025. In addition, fresh water supplies, essential for modern high-input irrigation agriculture, are becoming limited by increased human and agricultural use, and polluted by agricultural run-off and widespread use of agrochemicals.

It is feared that the resulting food shortages in the overpopulated parts of the world during the 21st century may lead to widespread social, economic, and political unrest, making food security the single-most serious threat to international peace and security.

At the same time it is well known that countries with efficient agricultural systems generally have high living standards, strong economies, lower rates of population growth and democratic forms of government Increasing food productivity in a sustainable manner will, therefore, not only provide adequate nutrition to the expanding humanity, but will reduce population growth, protect the environment, promote economic development and ensure social and political stability.

The challenge for the agricultural sector during the next few decades is therefore clear double food production by 2025, and triple it by 2050, on less per capita land, with less water, under increasingly challenging environmental conditions. The situation is further complicated by the fact that in spite of the heavy use of agrochemicals, modern agriculture still loses nearly 42% of crop productivity to competition with weeds and to pests and pathogens, and an additional 10-30% to post-harvest losses due to a variety of factors, especially in the developing countries where storage conditions are poor.

During the 20th century traditional plant breeding has brought about enormous increases in crop productivity. However, plant improvement by hybridization is slow, and is restricted to a very small gene pool owing to natural barriers to crossability. Beginning in the early 1980s, advances in plant cell culture can genetic transformation have overcome these barriers by making it possible to transfer defined genes into all major food crops, including cereals, legumes, cassava, potato, and many vegetable and fruits.

The entire global gene pool-whether it be plant, animal, bacterial or viral-is available for utilization. The first genes that have been integrated into crop species provide resistance to non-selective and environment friendly herbicides, and many pests and pathogens. Increasingly large acreages of transgenic maize, soybean, potato, tomato and cotton are being commercially grown for human use and consumption. Carefully planned introduction of such crops on a worldwide scale would greatly help in reducing or even elimination the enormous crop losses attributed to weeds, pests and pathogens.

The use of such crops will also have a beneficial effect on the environment by significantly reducing the use of agrochemicals. Other genes for improving crop productivity, and manipulating starch/protein/oil quality and quantity, resistance to environmental stresses such as temperature and drought are also being isolated and studied. In the foreseeable future, these will be used to produce second generation transgenic crops.

It is clear that during the next few decades a wide variety of transgenic crops will become integrated into agricultural systems in the industrialized countries, Introduction of such crops into the developing countries, which need them most, will be slow and largely through the efforts of multinational biotechnology companies, because most of the developing countries currently lack the scientific and industrial infrastructure to develop and introduce these technologies into their agriculture. In the long term however, much of the increase in food production to meet the dual challenge of population growth and food demand must occur in the developing countries.

This will impact greatly on their overall economic development, which will help control the relentless increase in the population. It is critical, therefore that scientific and technical manpower and infrastructure be created in developing countries to take scientific and technical manpower and infrastructure be created in developing countries to take advantage of the remarkable advances in agricultural biotechnology. Assistance should be provided to the developing countries be international organizations such as the United Nations Educational Scienticfic and Cultural Organization (UNESCO), the Food and Agriculture Organization of the United Nations (FAO) and the World Bank through its many international agricultural research centres, like those in Mexico and the Philippines, where Green Revolution originated, International agriculture in the latter half of the 20th century was dominated by the Green Revolution, Considering the power and potential of agricultural biotechnology, it is likely that the Gene Revolution will dominate the agriculture of the 21st century.

Like the Green Revolution, it may help accelerate the rate of food productions, save lives from hunger, create livelihoods in rural households, save large tracts of land that would otherwise be needed for food production, reduce birth rates, ensure low food prices, stimulate broad-based economic growth, expand world trade, and help create a sustainable agricultural system for future generations.

Finally, it should be understood that although plant breeding has long been, and will continue in the future, to be indispensable for plant improvement, it must now be complemented and supplemented by molecular breeding and genetic transformation, in order to establish a sustainable agricultural system for the 21st century.

Animal Biotechnology

Animal Biotechnology has developed rapidly from the early 1980s when the first transgenic mice and first in vitro produced bovine embryos occurred. Today animal breeding companies are using marker assisted selection to provide earlier and improved selection of breeding animals. Gene mapping efforts are annually identifying a large number of candidate genes for testing of usefulness in marker assisted selection.

The propagation of genetically valuable or transgenic animals is enhanced by in vitro production of embryos. There are already more than a dozen commercial companies selling or producing for customers in vitro produced embryos. Cloning of embryos of cattle and sheep has been shown to be possible, but low efficiencies have thus far prevented commercial use.

Animal producing and genetic selection systems can be made more efficient by sexing the sperm used for in vitro production of embryos. The flow sorting of sperm cells to separate X from Y bearing cells has been successful in most species tested and in cattle and swine has resulted in offspring of the desired sex. While effective with in vitro embryos production, the sexed sperm are of less than normal viability and sperm numbers and not yet useful in artificial insemination or the freezing of sperm.

The artificial insemination and embryo transfer industries are today heavily dependent on use of cryopreserved sperm and embryos. However the cryopreservation of oocytes and the combination of some biotechnologies such as embryo biopsy and sexing with cryopreservation are still in the research stage. Transgenic cattle, sheep swine and poultry have been produced primarily by microinjections of DNA into pronuclei of eggs or by use of viral vectors.

The intended use has been to change animal growth, disease resistance, wool production, or to produce new products in milk. The efficiencies thus far have been much lower than in mice. Therefore there has been little commercial use of transgenic technology except for the involvement of several companies in production of pharmaceutical products in milk.

The use of cultural embryonic stem cells and nuclear transfer to make offspring has resulted in calves and lambs. This technology offers promise of making large numbers of nearly identical offspring or more efficient gene transfer or selection. However thus far the technology is not sufficiently efficient for commercial use. Growth hormones and growth promoting factors are being developed for improved efficiency of animal growth. Growth hormones are commercially in use at present for promotion of milk production in cattle.

The anatomy and physiology of birds and the late stage of egg laying prevents the introduction of DNA into pronuclei as in manuals. Thus other methods such as oocytle microinjection of DNA, viral introduction of DNA and introduction of DNA into primordial germ cells are being used in avian gene transfer research. Oocyte microinjection, blastoderm transfection and primordial germ cell manipulation have all been successful in producing transgenic birds. The production of transgenic birds appears possible but is not yet in commercial use.

Patented applications of recombinant DNA biotechnology to animal agriculture became apparent in the early to mid 1980s with development of recombinant vaccines effective against calf scours and against rabies. Today numerous vaccines in use are produced by recombinant technology. Biotechnology has also revolutionized diagnostic techniques by introducing higher sensitivity and specificity. Monoclonal antibodies, DNA probes, Southern and Western blotting, DNA and RNA amplification through PCR, constitute as integral part of the diagnostic arsenal at the disposal of the advanced laboratories.

Based on these techniques rapid animal-side field test used by the clinicians furnish results in real time, permitting immediate decisions as to the proper treatment, vaccination, or stamping-out policy. High technology based diagnosis is expected to become a routine in every diagnostic laboratory forming a sound basis for control and treatment of animal diseases.

In the short time of approximately fifteen years, animal biotechnology has moved from an unknown and unstudied science to a strong discipline of science and extensive commercial application. The science and its commercial use are rapidly progressing. The ultimate benefactor will be humans through improved food supply, nutrition and health care products available to humans around the world.

Biotechnology in Agriculture

During the last decade, tremendous progress has been made in the area of plant cell, tissue & organ culture. In vitro techniques constitute an important component of biotechnology, and have the potential not only to improve the existing cultivars, but also for the synthesis of novel plants & early release of high-yielding plants resistant to various diseases, pests, stresses & temperature.

Green revolution of recent years has been the result of efforts that led to the improvement & better selection of species of agronomic interest. Some specialists feel that biotechnology is the second phase of green revolution, studies are already in progress to obtain crop plants having higher photosynthetic efficiency, improved fixation of atmospheric nitrogen, increased nutritional quality & greater tolerance of salinity, alkalinity, acidity, deficiency of nutrients & stresses.

Wide hybridization through embryo & ovule culture both at the interspecific & intergeneric levels have been achieved in a number of crops through the culture of the hybrid ovules & embryos. In vitro fertilization techniques have also been employed to overcome incompatibilty. Haploids produced through anther culture & their diplodization have resulted in the release of varieties in crops such as wheat & rice.

Biotechnology

The term biotechnology is composed of two words-bio (Greek-bios-means life) and technology (Greek technologia-means systematic treatment). In other words, it is the science of applied biological processes. Accordingly to the oxford Dictionary, technology is the "Scientific study of the practical or industrial arts". Spinks (1980) defined biotechnology as, "the application of biological organisms, systems or processes to manufacturing & service industries".

European Federation of Biotechnology (1981) defined biotechnology as, "the integrated use of biochemistry, microbiology & engineering sciences in order to achieve technological (industrial)application of the capabilities of micro-organisms, cultured tissue cells & parts thereof". In the report Biotechnology: A Development plan for Canada (1981) biotechnology is defined as "the utilization of a biological process, be it microbial, plant or animal cells, and their constituents, to provide goods & services".

Production of Pathogen-free Plants.

Plants traditionally being vegetatively propogated are systemically infected with viruses & other pathogens which greatly reduce yield & also Q uality of the marketed commodity. Virus diseases like potato leaf roll virus (PLRV) or potato virus Y (PVY) for example, can cause upto 95 % reductions in the tuber yield of potato crops. Increase in yield upto 300% (Averaging 30%) has been reported following replacement of virus-infected stock with specific pathogen free plants. Since majority of viruses infect plants in a systemic manner their elimination may be achieved through meristem tip culture. Plant species for which virus-free plants have been obtained.

Species Virus Eliminated

1. Allium sativum (Liliaceae) Garlic mosaic virus
2. Brassica Oleracea (Brassicaceae) Cabbage black ringspot virus, Turnip mosiac virus, cauliflower mosiac virus
3. Chrysanthamum sp. (Asteraceae) Chlorotic mottle, Green Flower, stunt vein mottle, virus B.
4. Dahlia Sp. (Asteraceae) Dahlia mosaic virus, Tomato Aspermy, vein mottle, virus B.
5. Glycine max (Fabaceae) Soyabean mosaic virus. Impomoea batatas Feathery mottle, Hanmon Mosiac, (Convolvulaceae) ' Rugosa mosaic
7. Lilium Sp. (Liliaceae) Cucumber mosia virus
8. Monihot Sp. (Euphorbiaceae) African cassava mosiac
9. Musa sp. (Musaceae) Cucumber mosiac virus
10. Nicotiana tobacum (solanaceae) tobacco mosaic virus
11. solanum tuberosum(solanaceae) potato virus.

Although the apical meristems are often free of viruses, this cannot be regarded as a phenomenon of u niversal occurrence. The success of meristem tip culture depends upon several factors. One of the most important factors is the relative distribution of viruses in the growing tip of donor plants. There are some viruses which invade the growing tip viz, TMV, potato virus X, cucumber mosaic virus. In such cases virus free plants are obtained by combining meristem-tip culture with high temperature treatments.

Production of Disease Resistant Plants

This is one of the most useful applications of tissue culture in crop improvement. In potato, somaclones have been screened for both late & early blight resistance. In maize, somaclonal variation has induced resistance to race T of southern corn leaf blighrt. In sugarcane, resistance to diseases like fiji & downy mildew have been recovered. In lucerne, selection of cell lines & plants resistant to the toxin of *Fusarium oxysporium* has been accomplished.

Improvement of Nutritional Quality

One of the major sources of protein for human & animal commnsumption is constituted by the proteins contained in seeds of many plant species. The cereals & legumes which are major sources of storage seed proteins, contain limited amount of certain amino acids which are essential for human beings. Majority of these cereals are deficient in lysine whereas legumes are deficient in sulpher amino acids. A wide range of approaches have been employed for improving nutritional quality of various crop plants. Important among them are selecting cell lines resistant to amino acid or their anologues of lysine, tryptophan, proline & phenylalanine. Isolation of variants over producing specific amino acids in culture has been successful, but expression in the whole plant & especially in the seed has not yet reached the level required to make an impact on protein quality.

Selection for Salt & Draught Tolerance

Continued efforts to increase intensity of cropping for increasing production from limited land resource by extending irrigation facilities have resulted in the gradual build-up of salt concentrations in the soil. This has resulted in loss of productivity on such soils. Salt tolerant lines have been produced in crop plants such as tobacco, tomato, cereals. Salt tolerance has been incorporated into rice lines with improved plant type. Some of the improved rice cultivars viz., IR 42, IR 43 & IR 52 are salt tolerant.

Development of cultivars tolerant draught can contribute significantly in agriculture economy. Tolerance to draught is a polygenic trait & involves highly complex osmo-regulatory functions. In tissue cultures, simulated draught conditions have been achieved through incorporation of non-penetrating osmotic solutes such as PEG (polyethylene glycol) and dextrans in the media.

Production of Genetically Variable Plants.

The success of any crop improvement programme depends on the usable genetic variability in the base population cells in culture offer an excellent systems for inducing variations & regenerating pure mutant types. Genetic variation can be an option to lessen our reliance on cost intensive germ-plasm collection & conservation programmes.

Variant producing capacity of cell culture can be augmented to a great extent by employing physical & chemical mutagens. Somaclonal variation has been extensively exploited for the improvement of a sexually propogated crops viz., potato, sugarcane. From cell cultures, some superior cultivars have already been produced in sugarcane which are high yielding, drought resistant & temperature tolerant.

Biofertilization

Molecular nitrogen in the atmosphere is converted into Biologically converted forms by nitrogen fixing micro-organisms e.g. Rhizobium. The most sophasticated approach to biofertilization is to create plants that possess the genetic capacity for nitrogen fixation. Attemps are being made to transfer genes for nitrogen fixation (hifgenes) from bacteria to plants.

Rapid Clonal Propagation

Tissue culture has found its best commercial application in production of cloned plants at a very high rate as compared to conventional methods. It is important specially for initially building up of propagation stock of elite clones or individual plants which are otherwise slow to multiply. A number of agriculturally important plants have been clonally multiplied.

Genmplasm Storage

The primitive cultivars & wild relatives of crop plants constitute a pool of genetic diversity which is invaluable for further breeding programmes. There are over 20,000 plant spices which are rare or threatened with degradation of their neutral habitats. The most economical form of storing germ plasm for seed propagated species is seeds. However, there are certain limitations of this method. Therefore, various methods of invitor storage of germplasm are thus of great practical significance for long term storage of germplasm. Presently, there are two approaches to invitor germ plasm storage. Slow growth technique & cryopreservation.

Biological Control of Agricultural Pests

Insects consume nearly one-third of human food supplies on earth. One of the major goals of biotechnology is to develop target specific biological pesticides that will kill some specific pest but will not harm other species. Several insect-specific pathogens, developed through biotechnological processes, are being produced commercially for their use as microbial pestices. *Bacillus thuringienis* inocula have been used as bio-insecticides. Toxin produced by the bactenium kills the gypsy moth. The prevention of crown gall disease formation is achieved by spraying young plants or seeds avirulent species *Agrobacterium radiobactor* var, *radiobactor.* Microbial insecticides have a number of advantages over traditional chemical insecticides:

a) they are specific for a small number of species of insects & do not kill plants, animals & beneficial insects.

b) they are cheaper than organic pesticides and

c) they do not leave any residue effect.

Selection for Herbicide Resistance

Herbicides are used to control weeds in agricultural fields. Tissue culture techniques offer several advantages in developing herbicide-resistance plants:

1. The techniques for selection are relatively rapid;
2. Small volumes of cells represent the genetic potential of hectares of fields.
3. Metabolic studies are facilitated by uniform & sterile conditions &3
4. Possible new types of resistance not found in nature may be obtained from plant cell cultures.

Attempts to produce herbicide tolerance using somaclonal variation & invitor selection have already been made.

3

Biotechnology in Agricultural Development

Introduction

We are at the threshold of 21st century, where great challenges on three fronts namely, production, pollution and population are awaiting the developing countries. Moreover, in the era of privatization, globalization and liberalization, there is heavy international competition within as well as between the countries on all fronts including the agriculture. The present scenario indicates that there would be "survival of the fittest". Needless to mention that by 2000 AD our population will be around one billion the priority is clear-cut before us. We have to climb upward to achieve higher and higher production to ensure that every mouth of our ever increasing population has enough to eat.

The quest for agriculturists for greater productivity and improvement of existing cultivars with better quality food continues with great vigour and spirit. The conventional methods of plant breeding and traditional agricultural practices have done a tremendous job and contributed to a great deal towards the above goal. However, in view of the acuteness of the problem and renewed fears regarding the availability of the proper and enough food, these methods alone are not sufficient to meet the situation.

Although, the FAO believes that food production will continue to increase for the next 20 to 25 years at the same rate as experienced during the past 30 years to meet the futuristic demand. The worldwatch Institute has, however, reported that this rate may not be sustainable. Moreover, they believe that science and technology can no longer ensure the onward march of achieving higher and higher yields. In addition we are also subjected to an almost daily litany of doom and despair as a result of global warming, depletion of ozone layer, oil and water running out and so on. The next century certainly packed with trouble if the meteor does not get us first, then we will gradually either be roasted, frozen or desiccated, and to make doubly sure, shall certainly be starved as well.

It seems that if we have to stand any chance whatsoever in the next century, our capacity to adopt the living resources will be extremely important for our survival. Because of changes in the environment and depleting energy resources, this capacity to adapt will be every bit as important to those areas of the world which are blessed with overproduction like Europe today. Accelerated adaptation through genetic change is, of course, one of the important means by which plant breeders have achieved their aims in the past. Today the new techniques of biotechnology in general and genetic engineering in particular may offer to the plant breeders and agriculturists the chance of speeding up adoption to an extent hitherto considered impossible. This new power will undoubtedly bring risks but it could also be the only chance we have for escaping the dangers ahead. Biotechnology being a multidisciplinary subject requires coordinated efforts and expertise of scientists with different backgrounds to put together in order to achieve success. It may be defined as may technique that uses living organisms, or parts of organisms to make or modify products, to improve plants or animals or to develop microorganism for specific uses.

It includes well established technologies such as those used in breeding, plant propagation and conventional animal vaccine productions. Modern biotechnology encompasses the more recently developed techniques involving the use of recombinant DNA technology, monoclonal antibodies and new techniques of cell culture. Recent development in biotechnology have made it possible to move genes from microbes, plants and animal species into plants of interest. Biotechnology helps in the commercial utilization of gene and other things for the benefit of human kind. However, destruction of ecosystems has been a major concern of biotechnology.

Developing countries are naturally attracted to the potential applications of biotechnological research in solving problems of hunger, energy supply and improving the quality of life. The priorities of different countries however very widely, In this context, the National Biotechnology Board of India has chosen genetic engineering, photosynthesis, tissue culture, enzyme engineering, alcohol fermentation and immuno-technology as areas of immediate interest.

Application of Cell Biotechnology

Cell biotechnology or plant tissue culture is an enabling technology from which many novel tools have been developed to assist plant breeders. These tools can be used to increased the efficiency of the breeding process, to improve the accessibility of existing germplasm and to create new variation for crop improvement. They include micropropagation, another culture, in vitro selection, embryo rescue, somaclonal variation, somatic bybridization and transformation. Of these somaclonal variation occupies a somewhat unique position, because it has both the advantages and disadvantages of tissue culture systems.

Somaclonal variation was first defined as such by Larkin and Scowcroft who reviewed the subject in 1981 and were among several authors at that time to draw attention to its potential use for crop improvement (Larkin and Scowcroft, 1981). An extensive number of reports soon followed in a whole range of species, indicating that somaclonal variation was widespread, and therefore, ostensibly accessible to all plant breeders (Karp, 1991).

Table : List of varieties developed from somaclonal variation.

Crop	*Country and Institution*	*New trait*
Geranium	Deptt. of Hort, Purdue	
	Univ USA	Vigour and attractive flowers
Sweet potato	North Carolina Res, Services,	
quality of roots	USA	Colour, shape and backing
Sugarcane	Sugar Research Centre, Fiji	Yield and disease resistance
Maize	Molecular Genetics, USA	Trytophan content
Tomato	DNA Pl. Tech. of	Dry matter content, Disease
	New Jersey, USA	resistance
Rice	Plant Research Institute,	
	Japan Unv of Agri. Sci.,	
	Godollo, Hungary	Yield Disease resistance
Celery	DNA Pl. Tech. of	
	New Jersey, USA	Processing, Yield & efficiency
Brown mustard	ICAR, New Delhi	Yield.

The Utilization of new genetic variability has become one of the major objectives of tissue culture, The assembly of genetic variability is vital for improvement of crop plants. Somaclones for the resistance of downy mildew. Fiji disease and eye spot disease were identifies. Application of somaclonal variation in crop improvement can produce increased yield and resistance to biotic as well as abiotic a stresses. Some of the crops where somaclones were produced are rice, wheat, maize, tomato, geranium, sweet pototo, sugarcane, rice, celery and brown mustard (Table 1). Our country research efforts have developed a variety called 'Pusa Jaikisan' in brown mustard evolved by somaclonal variation.

Micropropagation is another technique of great significance to agriculture which is based on in vitro culture of plant cell. Each and every cell of a plant has the potency to be regenerated into complete plant by tissue culture techniques and this property of the cell is known as totipotency. Using this technique hundreds and thousands of seedlings can be generated in a very short time, starting from a limited number of explants. This methods of multiplication has two major advantages. First is the multiplication of new elite cultivars. For example, after its release, a new sugarcane cultivar may take 5-10 years before sufficient planting material may be generated for wide coverage in the farmers field. But using micropropagation techniques a comparable level of multicipation may be achieved within 2 years. The second advantage is that we get disease free planting material because of the use of aseptic conditions during tissue culture. There is also reports of rejuvenation and increased vigous following tissue culture. This can also help in rescue of important materials which are on the verge of extinction due to disease attack. For example, a rare variety of banana used for temple offering in Karnataka has been rescued through tissue culture techniques. Micropropogation of fruit and ornamental plants provides a useful avenue for the employment generation for rural youths at cottage level industries. In species like Daffodies and gladioli, microporpagation techniques are being used to speed up the release of new varieties. In strawberry, millions of plants can be produced from a single mother plant in one year.

Micropropagation work undertaken at our Defence Agricultural Research Laboratory. Pithograh was resulted into successful multiplication of strawberry, petunia, carnation and tomato hybrid. In case of strawberry a well adapted variety of hill vi., Jyolikit Selection was taken for micropropagation studies with objective to large scale production of disease free propagules since this fruit plant is highly susceptible to number of fungal, bacterial and viral diseases. The result showed that internodal segments are most suitable explant material for micropropagation under in Vitro condition. MS formulated with 1ppm of BAP and 0.1ppm of GA3 was found optimal protocal for maximizing shoot proliferation in explant where MS media modified with 1 ppm 1AA was most suitable for early rooting in shootlets. Peat moss was found to be most suitable potting culture for hardening of tissue culture raised plants of strawberry var. Jyolikot Selection.

Micropropagation studies in Petunia Hybrida were undertaken in order to raise the plant seedlings in mass. The result showed that optimum number of shootlets could be obtained after culturing the leaf explant of size 5mm X 2.5mm in MS media supplemented with 1ppm of BAP + 0.5 ppm of IAA with an inclubation period of 34 days. Protocol standardisation revealed that MS medium supplemented with 1 ppm of IBA was suitable for an early and healthy rooting in tissue culture raised shootlets. Hardening process was most successful with high relative humidity (90-95%), Temperature 23 + 10 C and potting mixture of sand and soil in equal proportion.

Work on micropropagation of carnation was resulted into standardization of an vitro culture media for shoot and root proliferation. Protocol found most promising for shoot proliferation was MS medium supplemented with 1 ppm of BAp AND 0.1

ppm of IAA where MS medium supplemented with 1 ppm of NAA was identified to most suitable for rooting in shootlets.

Micropropagation studies were also undertaken in hybrid tomato to generate large scale seedings to fulfil the need of army personnel as well as local growers with pure and healthy planting material. Based on the observation it was found that tissue culture medium comprised of MS,0.5 ppm kinetin, 0.5 ppm IAA and 0.3 ppm GA3 was most suitable for shoot growth and MS medium supplemented with ppm of IAA or MS with 1 ppm of IBA was most suitable for early and healthy rooting in the explant material. Tissue culture regenerated plantlets showed high survival under relative humidity varied from 85-95 per cent, temperature 25+20C and potting mixture sand and oak forest humus in equal ratio.

Application of DNA Technology

Genetic engineering opens a totally new dimension for bioprospecting. The search for new genes and their application is the primary objective of the biotech industry. Gene technology now enable humans to integrate revolutionary new properties in to cultivated plants through inter-specific or inter generic gene transfer which was not possible through classical approach of crop improvement.

Genetic engineering is the direct introduction into a plant of an isolated or modified single gene using transformation techniques. Now it is possible to transform almost all of the plants cultivated by man. Where this has not happened, it is more to do with the insignificance of the plant in agriculture or forestry rather than a reflection of a stubborn resistance to all attempts to transform it. A couple of years ago, cereals were regarded as being recalcitrant species. This has new changed and transformation has been reported as being successful for nearly all cereal species.

Transgenic Plants

Transgenic is used to describe plants which have had DNA introduced into them by means other than by the transfer of DNA from a sperm cell to an egg. Development achieved through modern biotechnology in the form of transgenics have given new options of controlling insects, diseases and weed pests, while improving the overall integrity and consequences of agricultural practices. Biotechnology combined with modern crop management techniques and the responsible use of pesticides gives the best tools available to ensure healthy, high yielding crops. However, the benefit expected from the release of genetically modified organisms into environment are quite more which are seems to pay major role in bioremediation, environmental improvement, agriculture, food industry and health care. A number of transgenic plants in various crop species covering wide range of altered characteristics have been approved for commercial cultivation.

Resistant to Biotic Stresses

Damage to crops by biotic factors like insects-pests and weeds is a major limiting factor in agriculture economic in tropical and temperate regions of the world. Despite large scale investment in the chemical control of pests, they cause great economic loss by damaging or destroying the crops.

With a tremendous development in techniques to engineer crops genetically, we now have the ability to make broader use of natural insecticides. The choice of a Bt endotoxin, as the first insecticidal protein for introduction into plants, was based on the extensive knowledge gathered about this class of crystal protein since 1902 and such a strategy has been successfully achieved. The application of glyphosate kills crop plants just as effectively as it kills weeds. The usefulness of glyphosate as a weed control agent in agriculture could be enhanced if resistance to herbicide could be selectively engineered into crop plants. Transgenics have been developed by introducing the corresponding resistance genes from weeds and microbes into crop plants such as tobacco. In addition genes conferring resistance against herbicides, viz., Sulphobylurea, Imidazolinones, Phosphinothricin, Asulam, Bromoxynil and 2,4-D have been engineered and transgenic plants developed.

Abiotic Resistance

A large area of our country is under stress conditions i.e. saling, alkaline water, cold and heat stress etc. transgenics produced from introducing mtl 1D gene from E. coli in tobacco and arabidopsis showed tolerance to high leave of NaCl. Thermoc tolerance was found in Arabidopsis plants engineered with hsf gene. It has been reported that over-expression of sac B gene from Bacillus subtillis leads to high level of fructans in tobacco cells, and this is associated with increased drought tolerance whereas chemically synthesized antifreeze protein genelala 3 leads to improved freezing tolerance of tobacco plants.

Genetic engineering for improved tolerance to abiotic stresses is, therefore, the need of the hour as the existing cultivars in most cases are capable of giving much higher biomass that what we harvest today. While it is true that the response of plants to these stresses is multigenic, the recent success achieved in genetic engineering against excess salt, high and low temperature as well as less or excess water by altering individual gene is noteworthy.

Product Quality

A group of scientists have reported the ethylene synthesis in transgenic tomato plants, using anti-sense to a gene that has been identified as ACC oxidase. The enzyme convert ACC to ethylene. Ethylene production was inhibited by 97% with a signification reduction in over-ripening and shrivelling. Flavr Savr and Endless Summer in tomato, Freedom II in squash, high lauric acid reposed (Canola) and

Round Up Ready soybean are some examples of the crops that already being commercially grown in developed countries.

Genetic engineering of metabolic pathways of fruits and vegetables has therefore given control over post-harvest process which not only leads to improvement of quality characteristics but also enhances processability, transportability and prologation of shelf like.

In addition, engineered plants with change fatty acid composition of edible oil, reduced antinutritional components in many food crops, production of palm oil or animal derived industrial oil into plants like Brassica are some of the good examples of novel products that can be expected from biotechnology. Significant progress has been achieved for production of biodegradable plastics derived from biomsass of genetically altered plants having gene from bacteria.

With the advancement in DNA technology and understanding the structure and function of gene and its products more and more transgenics are coming out from various research programme all over the world. However, boon of biotechnology has been confined to few developed countries and developing countries are still in vogue to harvest the benefits of modern technology. More that 48 transgenic has picked up remarkably in last couple of years. During 1986-87, 25000 field trials were conducted with transgenics of more that 60 crops in 45 countries. Sixty per cent of these trials were in first 10 years and 40% in the last two years. Area under transgenics cultivation has also increased tremendously. During 1996, the transgenic crops covered an area of 2.8 million has which increased about 4 times during 1997 and 10 times during 1998 (Table 3)

Table : Area of transgenic crops planted (m ha)

Country	*1997*	*1998*
USA	8.1	20.5
Argentina	1.4	4.3
Canada	1.3	2.8
Australia	0.1	0.1
Mexico	<0.1	0.1
Spain	0.0	<0.1
France	0.0	<0.1
South Africa	0.0	<0.1
Total	11.0	27.8

Source: Anne Simon Maffat (1998)

A major challenge for agriculture in the 21st century in developing nations would be to speed up production process economically to fulfil the aspirations of huge populace, to achieve diversification and adding value to the primary produce so as to make agriculture enterprise farmers as well as environmental friendly. Biotechnology based advance technologies are expected to materialize many of our expectations in the coming millennium.

Viruses

To minimize the losses due to viral disease and to reduce the use of chemicals, the cross protection phenomenon of viral coat protein and movement proteins have been employed. This method has been used to develop transgenic plants which conferred significant levels of resistance in tobacco plants to TMV or to alfa mosaic virus (AMV) using the AMV coat protein gene. In some viruses it has been found that certain sequences termed satellite RNAs act to reduce disease symptoms in infected plants and as such it is being utilized for developing transgenic with virus resistance. Antisense RNA which binds to sense RNA and artificial ribozyme molecules which cleave specific gene transcripts were found very powerful tools for the inhibition of viral disease development. Genetic engineering of antisense or ribozyme genes complementary to viral genes whose products are essential may prove one of the best way of viral disease control.

Insects

Transgenic crop plants that produce pesticidal proteins such as Bt toxin can offer the growers advantage of increased yields, numerous environmental benefits due to decreased use of conventional insecticidals. A transformed tobacco plants with the gene encoding the cowpea trypsin inhibito, in feeding experiments demonstrated that expression of the gene conferred resistance to a range of insect genera, including Heliothis, Spodoptera, Diabrotica and Tribolium the protein is not however toxic to humans.

Herbicides

The use of herbicidea has become an important feature of modern agriculture as a means of controlling weed species, however, it is economically feasible to produce specific herbicides for use with every individual crop species. A more generic approach would be to transform a range of crop plants with a gene which confers tolerance to a single non-selective herbicide.

Biotechnology for the Cotton Farmer Problems and Prospects

Cotton or white gold as it is aptly called is grown for its lint and seed which yield cotton fibre and seed oil, respectively. This crop occupies 32-33 mha of world area with a production of 20-25 metric tonnes. In India its area spans over 7.5 mha with an average yield of 290 kg/ha of lint and 870 kg/ha of seed cottom. To meet the challenges of 2000AD with a population of more that 960 million, a total production of 19 million bales is required as against the 13-14 million bales of today. This can be achieved by the use of improved crop production practices coupled with appropriate pest management tactics. In addition, generation of novel. Transgenics may help achieve the near impossible. Genes that have been identified as potentially profitable, if engineered into acceptable cultivars can be used to generate such transgenics. Among these are genes imparting resistance to herbicides, insects, pathogens and abiotic stresses. It is also widely accepted now that a number of other qualitative characters can be improved, such as fibre strength, fineness, colour and thermal adaptability of the fibre.

Thansgenic plants have become realistic components of stress management world over. Bollworm and herbicide resistant transgenic cotton have received the approval of the Environmental Protection Agency and have been commercially released in the US for cultivation. Considering the fact that numerous biotic and abiotic stresses limit cotton production, it is likely that future strategies might orient towards the development of a multiadversity resistant high yielding transgenic cotton variety with superior fibre qualities.

The most important aspects in the development of transgenic plants are

* Identification, isolation and clining of the desirable genes
* Transformation of plant cells or tissues with a suitable vector
* Regeneration of the transformed cells or tissue
* Confirmation of the gene integration and expression
* Hardening and field establishment of the transgenics.

Major achievements worldover in the field of cotton transgenics.

Genes

Genes for Insect Resistance Used

* Toxin genes from Bacillus thuringiensis
* Protease inhibitor genes from plants & the horn worm

Genes used for herbicide resistance :

* Nitrilase gene from a bacteria Klebsiella spp. for Bromoxynil resistance
* Mono-oxygenase gene from a bacteria Alcaligenes eutrophus for 2,4-D resistance
* Acstolactate synthase gene for resistance against sufonyl urea & imidazoline group

* ERPSP synthase gene for glyphosate resistance.

Genes used to Improve fibre qualities :

* Polyhydroxybutyrate & polyhydroxyacetate from bacteria
* Expression of indigo pigment in fibres

Genes to be Used

* Cholesterol oxidase genes against insects
* Heliothis stunt virus genes against Heliothis
* Genes encoding chitinases glucanases, attacins cecropins vital coat proteins etc. against diseases.

Transformation

* Agrobacterium medicated transformation
* Particle acceleration gene delivery systems have been successfully used all over the world.

Regeneration

* Callus fissue regeneration-mostly genotype specific
* Protoplasts were regenerated successfully.

Meristematic tissues were successfully regenerated.

Expression

* Enhanced expression was achieved using truncated forms of the full length genes, repeat copies of the propoter, modification of the bacterial codons to plant prefered type.

Commercial Release :

* Bollworm resistant transgenics-NUCOTN 33M & NUCOTN 35B were released in the US and cotton transgenics-are in now under field cultivation in Australia, South Africa and China.
* 8 lakh hectares were planted with the transgenic cottons in 1996 and 1997
* More than 8,000 hectares were damaged by bollworms.
* Transgenic cotton plants in Australia were also found to be attacked by the bollworms.
* Herbicide (Bromoxynil) resistant transgenics BXN 57 and BXN 53 were released in the US in 1996.

Problems Encountered/Expected

* Gene silencing and expression instability

* Resistance in insects to the toxins used. Studies on baseline toxicity indicated that there is a natural variation in the ability of Helicoverpa armigera in tolerating the CrylAc toxins. Some strains are capable of surviving the totoxins by virtue of an in built capacity to tolerate the toxins, while others are not. While development of Helicoverpa resistance to the CrylAc toxins itself may not be of immediate concern, a slight shift in the tolerance levels resulting from continuous exposure to the transgenics, can cause control difficultes.
* High cost of transgenic seed.
* Microecological changes resulting in pest shifts. CrylAc is a broad spectrum lepidopteran toxins that can cause significant changes in almost all populations of lepidopteran insects occurring on the cotton crop. Some of these insects, which cause negligible economic damage to the crop, harbour parasitoids which have the potential to keep the bollworm populations under check. Consequently, in the absence or low populaions of natural enemies, the bollworms can emerge as stronger pests that before. Insects pests such as Spodoptera litura which are less affected by the CrylAc toxins and which have been under check due to the use of pyrethroids can resurgacc as major pests as pyrethroid use is likely to be reduced on transgenic Bt cottons.

Some Recent Development in Cotton Transgenics

* Three Bt transgenic lines, 95-1, RIOI and R-19 have been indigenously developed in China through Agrobacterium medicated (Somatic embryo) and pollen tube pathway transformation. Thestability of Bt gene expression and field efficacy have been confirmed.
* Microprojectile bombardment parameters for pollen mediated transformation and genetype independent protocols have been develop in China.
* A new methods of transformation through injection of Agrobacterium into developing embryos, has been reported from South Africa.
* Stable transformation and regermation has been reported for Bt transgenics in Uzbekistan, China, Egypt and Australia. Pakistan reports development of cotton transgenic plants resistant torearcurl virus.
* An antisense DNA of CLCUV DNA-A bome ACI gone along with the antesense DNA of the AC2 and AC3 gene was used for the vector construction and transgenic cotton resistant to the CLCUV was developed in Pakistan.
* In Russia, A glucanase gene has been isolated from thermo philic bacteria and is being used as a new reporter system fortransgenic cotton development.
* Genetically engineered Baculovirus ACMNPV with genes from straw itch mite has been developed been developed by Monsanto and Zeneca with researchers from Madison, USA.

* Bt rransgenic cotton currently occupies 20% (85,000 ha) of cotton area in Australia. Isecticide use has been reduced by about 60% The efficacy has been moderate. The transgenic plants were found to express Bt proteimns only till the 95th day after which fluctuaions were observed. The reduction in expression was primarily due to down regulation which post transcriptional changes of the unstable RNA It was also reported that cotton tannings which increase with growth phase, act as antagonistswith Bt toxins.
* After two years of field cultivation, true resistance to Bt has not yet been detected in Australia.
* Reslistanor to Bt in field populations of Helicoverpa armigera is reported form China
* Annually 100 million dollars are now being spent in the US, only in search of new insecticidal genes.
* In the US, 45% of cotton area this year will be under transgenic cotton (herbicide and insect resistant). However, the area under Bt cotton seems to on the decline.
* Transgenic cotton developed in Australia using the Stunt virus genes, is not performing well.
* Transgenic cotton developed in the US using lectin genes, is not performing well against Heliothis virscens.
* The Australian report that Helicoverpa armigera is at least 100 times less susceptible to Cry toxins as cinoared to Heliothis virescens.
* The Window strategy for insecticide resistance management in Australia is still important for transgenic cotton.
* Cotton Pest spectrum in the US is getting altered after the introduction of Bollgard. Unsprayed Bt cotton sustained 4 times more attack of tarnished bugs, 2.4 times more with boll weevil, 2.8 times more with stink bugs and Spodoptera.
* Due to these changes in pest complex farmers had to spray 3-5 times on boligard as compared to 6-8 times on non-Bt cottons.

Prospects and Potential

The application of biotechnology in cotton farming can be either in the form of production of fermentation products or novel recombinant products for use (example, recombinant Pseudomonas expressing CrylAc) or as transgenic plant with in buitt resistance to biotic and abiotic stresses. Transgenic crops with in built resistance to insect pests and diseases can be extremely useful as this would result in the reduction of insecticide use apart from making pest management simple for the farmer.

Introduction of the bollworm resistant transgenic cotton is expected to reduce the use of chemicals used to control bollworm especially Helicoverpa importantly at a time when resistance to most insecticides including pyrethrolids is on the increase all over the world. Even if the introduction of the Bt cotton in India could result in a 25-30% reduction in insecticide use on cotton, this would mean a benefit of about of about Rs 300 crores, apart from the f-avourable impact on the environment so far, transgenic plants have been produced in about 60 plant species. Cotton has received special attention of the biotechnological companies in the developed countries who were attracted by the profit motives associated with the high value added to the transgenic seeds. The firs Bt transgenic cotton has already been released, as Bollgard. The Delta and Pineland using DP 5410 and DP 5690 as recurrent parents. The D and PL brand BT transgenic were labelled as Nucotn 33B Two million acres were planted in 1996 in mid-south region of USA, mainly in regions where Heliothis virescens was problematic. Nearly 1.2 M acres were under Nucotn 33 B and the rest under Nucotn 35B.

The Bollgard brand Bt cotton seed was sold at US $34-36 as compared to US $8-9 per hectare for non-engineered cotton seed. The average cost of control of cotton insect pests in the US was approximately US $150 in the early 1990s. Hence the prices were still found to be attractive. The transgenics were found to be more effective against Helothis virescens as compared to pectinophora gossypiella and Heliothis zea.

Farmers growing transgenics had to sign'an agreement with D & PL stating that they would not keep seed for planting next year. Quick ELISA tests have been devised to test for Bt toxins in plant parts, to check the illegal spread of transgenics.

Two more transgenic cotton varieties tolerant to herbicide Bromoxynil, trademarked as BXN 57 and BXN 53 were developed from the recurrent paren Coker 315 BYTHE Stonevuille pedigree company in collaboration with Calgene Inc. This seed was priced nearly 1.7 times the straight varieties.

In Australia the CSIRO and Narrabri research centre, in collaboration with Monsanto have developed transgenic cotton plarts that were commercial released in 1997. This introduction resulted in 50-60 % reduction in the US $ 93 million spent by farmers each year on insecticides. Pakistan which uses 90% of its total insecticides on cotton alone and China which is one of the major consumers of insecticides on cotton are reportedly strongly considering the prospects of introduction of Bt transgenic cottons.

Cotton acreage in India is only 5% of the total cropped area, yet it consumes more than 50% of the total insecticides used in the country. The pattern of usage, however, is not uniformly, spread, for instance in Andhra Pradesh alone, where cotton cultivation is only 0.3% of the total cropped area of the country, accounts for 17% of the pesticide use on cotton in India. Again within the average picture of the state, coastal AP uses 30% more pesticide than the state average.

The approximate estimates of insecticide sales in Guntur district alone, where cotton is grown in 0.15 mHa, is about US $ M. Clearly areas with maximum pecticides use per hectare, necessitated mostly due to bollworms, are likely targeted market niches for the bollworm resistant transgenic cottons. Moreover, since the price of the transgneic seed in the US is nearly four times the high input cotton cultivators of the irrigated belt Growing Bt transgenics and herbicide tolerant transgenics on marginal land is not recommended by the company as these would not show any impact in the absence of strong insect pressure and broad leaved weeds. In most areas of rainfed cottons the use of insecticides and pest pressures are low hence, it may not be economical in such belts, to encourage the use of transgenics.

Now transgenic production technology for many crops has become an established routine procedure in many countries including India. The research on transgenic cotton in government funded research labs in India is almost in the final stages. Meawhuile Mahyco-Monsanto biotech is ready to market Bt transgenic cotton (bollgard) in India.

The introduction of transgenics through these companies are to be in the form of FL seeds wherein Bt genes are inherited from the transgenic exotic germplasm. Such seeds may also be priced at a premium and it remains to be seen how enthusiastic the response of Indian farmers would be to the expensive input. Transgenic releases from government organisations would definitely enforce a competitive pricing to restrict the monopoly of the private companies.

The transgenic technology is also likely to be extremely beneficial to the private sector. For instance, a transgenic with resistance to a specific herbicide marginalises the use of other herbicides for chemical weed control and the chemical company holding the patent on the herbicide might be prove to be beneficial to the resource rich farmers. Clearly, here the targeted market is primarily high input farmers. The monopoly of transgenics due to patenting rights held by the biotech companies may also lead to high pricing of the seeds.

A Significant socioeconomic issue that arises from the introduction of transgenics into the Indian farming system is that the high priced seeds may benefits the prosperous and large farmers thus providing a negative externality on small and marginal farmers. On the other had it is also argued that the developments from the application of biotechnology would be beneficial to low input farming practises wherein the cost of chemical inputs can be minimised. It is now only a matter of time before we experience the full social. economic and environmental impact of transgenics in our country.

Biotechnological Tools for Increasing Productivity of Pulse Crops

Pluses play a significant role in Indian agriculture because they provide protein rich diet to poor people of the country. They contain 20-29% protein i.e. about 2.5 times more than cereals. Inspite of release of large number of high yielding varieties, extension of irrigated areas and use of agrochemical, growth recorded in total production of pulses from 11.82 million tonnes in 1970-71 to 13.19 million tonnes in 1995-96 is considered to be non revolutionary. This resulted in decline in per capita availability of pulses to 37g/person/day. There are several reasons for stagnation in pulses production such as narrow genetic base, lack of genetic variability in useable genepool, inability to transfer increased input to increase output (seed yield) susceptibility to large number of biotic and biotic stresses, photoperiod sensitive and non synchronous flowering and maturity. It has been estimated that on account of insect pest complex infesting pulse crops alone, nearly 2.0 to 2.5 million tonnes of pulses is lost annually. The monitory value of this loss will be approximately Rs.3000 to 3750 crores.

Conventional methods of crop improvement have not improve productivity of pulses with significant pace due to several reasons. These methods are slow, expensive and time consuming. Besides lack of reliable resistant sources, undersirable linkages and long gestation period are considered to be the main limitation of these methods. Hence, there is dire need to introduce innovative biotechnological approaches to enhance pulse productivity to meet out the market demand.

Biotechnology and genetic engineering has emerged as potential tool in recent year for genetic manipulation of crop plants. The said technology promises to improve crop productivity through complementing traditional breeding by decreasing dependence on harmful chemicals pesticides. Fertilizers and antibiotics etc. The tools of biotechnology are now available to plant breeders and can be used effectively to supplement conventional practices in improvement of pulse crops. It is expected that products of biotechnology will be ecologically sage, economically feasible and technologically expectable to Indian farmers.

Effort have been successful to develop several intergeneric and interspecific hybrids using embryo/ovule/ovary/endosperm culture in grain legumes, Several useful genes imparting resistance to various biotic and biotic stresses have been successfully introgressed on wild relatives of lathyrus, lentil, chickpea, pigeonpea urd bean and mung bean. The resistance to Aschochyta blight to chickpea has been successfully transformed from wild species Cicer judaicum to elite chickpea genotype PDG 84-10 at IIPR, Kanpur.

The techniques of in vitro mutagenesis in vitro selection and somaclonal variation are being also successfully employed to various pulse crops to develop promising genotypes resistant to various biotic and a biotic stresses. This technique has resulted success in developing Aschochyta blight resistance in chickpea. Further early results in vitro selection for salinity tolerance in chickpea is also very encouraging at IIPR, Kanpur. Similarly, the technology of cell suspension culture followed by selection of mutants has also been used successfully in major pulse crops at different centres in the country.

Identification and characterization of pathogens and pests is problematic and always pose challenge to plant pathologist and breeders in varietal development of pulse crops. The techniques of molecular diagnosis and DNA finger printing has facilitated resulted easy diagnosis and the precise characterization of pathogeny isolates and insect biotypes at molecular level. This has increased the efficiency of breeding programmes. Important pathogent like Aschochyta rabi and Fusarium oxisporum of chickpea have already been characterized using DNA fingerprinting techniques.

In a new world order and changed senario the characterization of germplasm and plant material is very essential to protect out natural wealth. The techniques of DNA fingreprinting have been proved very handy in protecting the intellectual property rights. These techniques are now being successfully used in pulse crops to characterize and catalogue the accessions of gene bank to protect the genetic wealth.

DNA marker techniques are the new innovations and have tremendous potential in increasing the efficiency of conventional breeding programmes. DNA marker technique can be used for tagging important genes which can subsequently be used for early, easy and precise selection in segregating of promising plants populations. DNA markers have been found linked to wilt resistance in chickpea, powdery mildew resistance in peas and nemarode resistance in mung bean and urd bean.

Transgenic technology which involves insertion of foreign DNA sequence has tremendous potential for improvement of crop plants. Efforts are on to develop transgenic plants resistance to pod borer in chickpea and pigeroxpea using Bt. Crystal protein gene from a soil bacterium. Early results obtained at various centres in the country are very encouraging. Similarly, efforts are also being made to develop transgenic urdbean and mung bean using coat protein gene. Current status and future strategies for genetic manipulation of pulse crops using biotechnological tools will be discussed in detail.

Biotechnological Applications for the Improvement of Indian Tobacco

India is one of the top 10 countries cultivation tobacco for the part ten decades both for domestic and international market, India is the third largest producer of tobacco in the world with an output of approximately 520 m kg. Of which FCV tobacco account for 120 m kg. The economic importance of to Indian exchequer is peogressively increasing in the last 10 years. Tobacco generates 10% of total excise revenue and tobacco 4% of the value of total agricultural exports, India has the unique distinction of growing different types of tobacco in different Agroclimatic zones.

Tobacco provides direct and indirect lively-hood to 26 million in India including 6 million farmers and workers. Many small and marginal farmers were benefited by growing tobacco. Research efforts to identify alternative crops to tobacco indicated that no other crops is as remunerative as tobacco. Stability of the worldwide demand of the crop, the crops hardiness and ability to grow in climate and soils non-suitable for other crops, relative ease in transporting tobacco and tobacco's profitability compated to other crops are incentives for farmers growing tobacco thus making tobacco one of the most appealing cash crops to farmers.

Tobacco crops is considered as major employer worldwide. Despite technological progress tobacco cultivation remained as one of the most labour intensive crops among all other agricultural crops. Women in particular and skilled labourers, who would otherwise have little chance of employment are able to earn living by working in tobacco cultivation. WHO in its report has acknowledged that in developing countries tobacco can create vast employment. In view of its great role in the economy of the country and farmers, further improvement of tobacco crop through Biotechnological tools will benefits them a great deal.

Biotechnological applications: Scientific system of tobacco cultivation in India was introduced with the establishment of Central Tobacco Research Institute (CTRI) at Rajahmundry in 1947.

Tobacco research in India over the past 50 years through CTRI has made significant contributions to the economic upliftment of tobacco farmers by identifying high yielding varieties and suitable package of practices This is reflected in the good reputation for Indian Tobacco in the International market, as leaf of desired quality traits as per the specific requirement of traders is produced.

Initially, conventional breeding methods were utilised for improving the yield and quality of tobacco. However, later application of biotechnology tools for supplementing the conventional tobacco improvement programmes or initiated. Different biotech techniques like anther culture, fertilized ovule culture micropropagation, protoplast culture somatic hybridization and also genetic engineering are being utilised. These techniques are helpgul is strengthening the conventional breeding programmes where ever they are week and failing to serve the purpose. Progress of biotechnology world, in different aspects, at CTRI is given below:

Yield Improvement

Yield improvement through conventional breeding programmes takes fairly long time. This can be reduced substantially (3-4 years) by culturing anthers (Anther culture). Of elite lines from early breeding populations in tissue culture. Through anther culture several dihaploids are produced and evaluated every year.

An improved line D1 was developed, in this way, which is superior in yield. Efforts to incorporate budworm resistance through someclonal variation in this line is in progress. Stable resistant D1 someclones are there in field testing stage.

Resostance to Biotic and Abotic Stresses

Wild species of tobacco possesses resistance to various tobacco pests and disease. It is difficult and in some cases impossible to transfer this resistance from wild species to cultivated tobacco through conventional methods.

In such cases, inter-specific hybrids were rescued through fertilized ovule culture and mass multiplied for further testing by micropropagation, In this way, interspecific hybrids possessing resistance to leaf spot diseases, black shank, root knot nematodes leaf eating caterpillar, aphids, budworm etc. were developed and evaluated for their resistance. Thus, stable pest and disease resistant lines were identified through Biotechnological tools. Efforts to transfer desirable characters from wild species through somatic hybridization is in progress.

Somaclonal variation was attempted to fortigy tobacco lines resistance to white fly (Bemisia tabci Genn.) (leaf curl) and budworm (Heliothis armigera Hb) and the results are encouraging. Somaclonal variation is highly effective to create variation when desired variation is impossible to get through conventional breeding methods and also variation is required only for one character.

Transgenic Bt tobacco cultivars processing resistance to leaf eating caterpillar (Spodoptera litura F) and budworm (Heliothis armigera Hb) were developed. Production of transgenics will give scope for transferring desirable characters from different species and genera.

Preliminary attempts were made to screen lines for drought tolerance in tissue culture. Such efforts will save expenditure on extensive field evaluation and work can be carried out in the off season also. Transgenic tobacco having wheat desiccation tolerant protein were evaluated water stress tolerance.

Herbicide Resistance

Orabache cernua os a root parasite of tobacco causing heavy loss in yield and quality. Glyphosate at lower does was found to kill Orabanche, but it is phytotoxic to tobacco. Hence research on production of herbicide (glyphosate) resistant tobacco lines through tissue culture was initiated. When farmer grows herbicide resistant lines in the field he can use berbicides without any harmful effect on tobacco, for the effective control of weeds and Orabanche.

Conservation of germplasm: Germplasm is the valuable source for various characters that are useful in future breeding programmes. This can also be conserved under in vitro conditions for long time. Also lines that cannot be maintained under field conditions due to non-flowering and sterility can be effectively maintained under tissue culture conditions. Interspecific F1s, mammoth mutants, in vitro rescued hybrids protoclones, BT transgenics various mutants are routinely being maintained.

Future Thrusts

Biotechnology can help Indian tobacco farmer in many ways, Particularly in tobacco improvement research and seed purity disputes, if offers a large scope. Research on tobacco improvement benefit farmer as he can get high yielding and pest and disease resistant tobacco for cultivation.

In research various aspects like in vitro conservation and molecular characterisation of germplasm, isolation, characterisation and transfer of desirable genes from across genera, molecular aided selection in breeding programmes, someclonal variation can play a great role. Utilising Biotechnological techques tobacco lines that can yield higher amounts of phytochemicals which are useful in the production various drugs and pesticides can also be developed.

In recent times many disputes about seed purity among farmers are coming up. At present morphological criteria is only being used for varietal identification. As morphology is highly subjective in nature and also varies with the type of environment the plant is grown, it is becoming highly difficult to solve the disputes. Under this situation molecular characterisation of cultivated varieties will help to identify correct varieties beyond doubt.

In future, giving emphasis on the above said lines will go a long way in benefitting the India tobacco crop as well as the farmer. However, as biotechnology got wider applications, any thing that put the farmer and environment at disadvantage need to be discounrages.

Biotechnological Approaches to Weed Management

Weed management is essential for good quality and quantity of food production. The presence of weeds in general may reduce crop yield by about 30% or more. Being potential pests, weeds have to be removed employing mechanical, chemical cultural and biological means. Weed management through application of chemicals (herbiciedes) is, through effective, not eco-friendly. The herbicides may pollute the environment and many alter the natural equillibrium. They also endanger and alter plant and microbial biodiversity.

Effective non-chemical weed management may include various biotechnological approaches using plants and/or their products. These are useful both for the natural equilibrium and the environment. In addition they also condition the soil and improve crop production. Some of the recently evolved biocontrol and biotechnological methods of weed management developed and tried at the National Research Centre for Weed Science, Indian Council of Agricultural Research, Jabalpur are outlined in the following text for ready use for trials and for working our their cost effectiveness.

Management of Phalaris minor in wheat by using Fungi

At present Phalaris minor is one of the most predominant weeds in wheat fields. The Weed spreads through seeds. The weed has developed resistance against popular herbicides like isoproturon. Biotechnological approach using fungi offers an alternative possibility of management of the weed through biocontrol.

(i) Control of Phararis minor using Trichoderma viride: Trichoderma viride is used as a seed dressing fungi in various crops including in wheat for management of seed borne/soil borne pathogens. Interestingly, the fungus has been found to inhibit germination of seeds of Phalaris minor. Application of Trichoderma, viride grown in saw dust and neem oil cake as a base consistently inhibits germination of phalaris minor seeds and reduces vigour of the seedings of the weed. At the same time it improves yield of wheat by about 5%. The fungus has potential of controlling phalaris minor. Soil application of the fungus, wheat seed treatment and spray of the fungus were all effective in controlling the weed.

(ii) Control of Phalaris minor by gilocladium virens: Control of Phalaris minor by ?Gliocladium virens has also been found to have potential for control of Phalaris monor by inhibiting seed germination in a similar manner as does the Trichoderma viride.

Management of Parthenium (Parthenium HysterophorusL) by Marigold (Tagetes PatulaL)

Parthenium is an obnoxious weed of worldwide occurrence, harmful to human and animal health, agriculture, environment and the natural biodiversity. The weed reduces yields of crops and grasses. When shown in parthenium infested field or area, marigold inhibits seed germination and growth of the weed. It also reduces population of the weed. In the next generation, reduction of parthenium population is as height as 85-100%. This has been found primarily due to competition and through all elopathy (affecting parthenium plants by releasing chemical substances including thicophene into soil through roots.)

Built up of the phytotoxic substances of marigold origin prevent parthenium seed germination, growth of plants, and flower and seed production completely in 2-3 years. This technology appears to be economically rewarding for marisgold flowers have good market.

This can be achieved simply by spraying marigold seeds over parthenium infested area. The technology is simple, effective, economically rewarding, and ensures self perpetuating control of parthenium in wastelands. Marigold to parthenium plant ratio of 0.5 to 4 appear effective for near complete control of parthenium. Adequate moisture availability in soil facilitates establishment of marigold in parthenium infected area.

Management of Parthenium by fungi: Two fungal species namely Gliocladium virens and Trichoderma viride which have been widely used as seed dressing to present parthogenic damage have also been found effective for parthenium control.

(i) Control of parthenium by Gliocladium virus: Glioladium vieus and 10% neem oil separately as well as in combinations spray control pathenium and other broad leaf weeds like chenopondium album, Melilotus alba and Medicago sp. In wheat crop. The fungus could be explored as a self perp control measure for non-cropped area also.

(ii) Control of parthenium by Trichoderma viride and need oil: Thrichoderma viride and neem oil separately as well as in combinations spray control parthenium under field conditions. The fungi has potential of becoming a self perpetuating means for control of parthenium under non-cropped area. Trichoderma viride also acts as an antidote for parthogens of parthenium that can possibly be used as biocontrol agents. If any of the parthogens attack crops, the Trichoderma virid can be used to overcome the problem.

Management of Parthenium by Utilising as a Source of Nutrients

Parthenium is able to extract nutrients even from nutrient deficient soils in which it grows. It has very height levels of nitrogen (3%), phosphorus (0.2%), potassium (4.5.%) and other macro and micro-nutrients. Uprooted, dried and cut to small pieces or powdered parthenium plants at preflowering stage can be applied directly to crop fields. Apart from increasing crop growth and yields, the parthenium residue conditions soil with the organic supplement. Keeping safety considerations, it may serve as an incentive for participation of public in its ecofriengly control.

Weed management through mid-summer irrigation and covering the beds with black polythene: Weed problem in most cases is due to seeds of weeds present in soil seed bank which germinate before, with or after planting the frops. Mid summer irrigation and covering the beds or fields with black polythene thoughout the summer kill most of the readily geminable seeds in soil seed bank. Polythene covering can be removed after summer before planting the crops. The treatment may provide near complete weed control.

This technique is suitable for areas where summer temperature exceeds 400 c. Weed management through black polythene stripe mulching Where crop is grown in rows, inter row space could be covered with black polythene mulching. This kills weeds appearing in the area due to interception of light necessary for survival of the plants. The are many other ways of biotechnological management of weeds such as bioling water jet treatment of soil, by intensive cropping, by increasing plant density and chosing allelopathic crop species etc.

Biotechnologycal Application

Techniques such as embryo transfer, dilution of semen and its cryo preservation and artificial insemination (A.I.) are well developed and expertise in these fields is available. This technology is extensively used in the case of cows and to some extent Buffaloes. The main lacuna in the case of sheep and goats is the non availability of improved breeds which could be utilised. As these animals are owned by the poor strata of society and there is a mistaken prejudice that they are responsible for environmental degradation no sustained efforts have been made to improve indigenous breeds and thus make available genetic stock for upgrading. In spite of this the numbers of these animals have been increasing rapidly (*2.5% per year in the case of goats) due to an increasing demand for meat. Our breeds have developed over the centuries to survive and adapts themselves to harsh environmental conditions but do not universally possess ability to respond to better management and nutrition.

Outstanding work in the case of the Sirohi goat was carried out from 1981 to 1993 by the Indo-Swiss Goat Development and Fodder Production Project in Rajasthan where they showed that outstanding animals are available within this breed. However, before the impact of this project could be felt the project was discontinued.

Goats

Out institutes is a very small NGO with only three technically qualified staff, one of who is in an honorary capacity and one on deputation from the Government of Maharashtra. We decided to initially concentrate on crossbreeding using the outstanding South African Boer Goat the worlds only meat goat as an improve breed. With assistance from the Government of Australia we imported 20 embryos of this breed from Australia implanted them in local goats and now have a small flock of Boer goats.

This goat unlike many temperate breeds of Dairy goats have proved to be extremely hardy, producing excellent crossbreed when crossed with local goats. In order to utilise this germplasm effectively we have developed and modified the techniques of semen cryopreservation so that it can be done with low cost equipment producing pellets instead of straws.

We have also improvised. A.I,. techniques using cervical insemination to get better conception rates. We have so far trained over 50 trainees among them farmers. We have supplied pure Boer goat frozen semen to Governmental agencies in Orissa, Tamil Nadu as well as to may private goat rearers in Maharashtra.

Our veterinarian is capable of carrying out multiple ovulation and embyo transfer if there is need for rapid expansion of our flock. However, this technology is expensive and only to be used if costs justify its use.

A I technology in goats has been spread to near by villages where fresh diluted semen as well as frozen semen have been used to produce 800 crossbreeds over the past five years. This technology is viable and easily transferable but unfortunately dependent on the sincerity of the inseminator.

Sheep

We were unable to identify a suitable improve breed to be imported which had the most desirable but greatly neglected attribute of prolificacy. We did locate a prolific and unusual breed in the hot humid sundarbands of West Bangal called the Garole. Up to 1993 this breed had remained unnoticed by the community of sheep breeders. We did however import a small flock of Awassi Dairy Sheep from Israel. This breed has produced an outstanding crossbred animal when crossed with the Malpura of Rajasthan.

This crossbred has been developed by the Central Sheep and Wool Research Institute (CSWRI) using Awassi rams obtained from us. The Australian Centre for International Agricultural Research has sponsored a collaborative project between the University of New England, Australia, The national Chemical Laboratory, Pune and the Nimbkar Agricultural Research Institute (A SISTER NGO of Maharashtra Goat and sheep Research & Development Institute), Phaltan entitled," Prolific Worm Resistant Meat Sheep for Maharashtra, India.

This project is making full use of Biotechnology. We are comparing three breeds of Indian sheep the Grole, Deccani and the bannur. We hope to evaluate lamb production including prolificacy and resistance to parasites and try and determine the genetic basis of these differences by DNA analysis. We also hope to account for the interaction between measurement of resistance to parasitic nematodes and environmental effects. The project has completed one year and the results so far are encouraging.

Our ultimate objective is to provide shepherds with rams incorporating the prolificacy of the Garole, the growth potential of the Awassi and the meat quality and hardiness of the Bannur. A very important but at present neglected application of Biotechnology which requires considerable investment is in the field of disease diagnosis and vaccine preparation. Once productive animals that respond to inputs are available and the rearers can afford to pay for the above services I am confident that the private sector will enter this field to provide this input.

It is imperative to review the cost benefit ratios of Government institutions and their contribution to the Goat and Sheep industries I think you will find that funds given to NGO's are more productive. There is much talk of helping NGO's but real assistance only seems to come from foreign aid agencies which is unfortunate.

Bacterial Biotechnology

The Production of high quality agricultural products in quantities sufficient to feed the world's people is clearly a matter of vital importance. Any general improvement in agricultural practices can have a tremendous economic impact, and the methods of biotechnology suggest at least two important strategies for bringing about widespread improvements.

Improvement through the use of symbiotic or potentially pathogenic microorganisms- Plants live in intimate relationship with many microorganisms. Leaf surfaces are covered by layers of symbiotic & sometimes pathogenic bacterial & yeasts, and the soil close to the root area is populated by bacteria species very different from the species found in soil away from plants. There have been a number of attempts to modify symbiotic microorganisms so as to make them even more beneficial to plants.

Improvement through the production of transgenic plants- Plants now being cultivated represent the end result of many years of pain staking efforts at improvement. Traditional plant breeding programs rely on two methods; the selection of advantageous spontaneous mutations & the introduction of desirable traits from closely related species by crossbreeding. Recombinant DNA Technology is bringing revolutionary change to these endeavous. It enables scientists to select precisely defined genes with well-characterized properties, clone them, and transfer them to a given plant. These genes can be taken from organisms unrelated to the target plant-even from bacteria or animals.

Transferring cloned genes from prokaryotes (bacteria) into higher eukaryotes, such as plants, is difficult. However, there are two species of bacteria, Agrobacterium tumefaciens & A rhizogenes, that transfer a small piece of DNA into plants as part of their normal life cycle.

Use of symbionts & Pathogenes

There are many Symbiotic bacterial associated with specific organs of various plants, a simple plant might be to modify such bacterial & then use such interactions with the plants to introduce the modified traits at the appropriate locations. This alternative approach is technically easier, because the engineering of bacterial DNA through recombinant DNA methods is now routine, whereas the manupulation of plant DNA is yet to be perfected.

(a) Protection of plants from frost damage via engineered symbiotic bacteria: One of the earliest examples of successful modification of a symbiotic bacterium was performed in Pseudomonas syringae, which is found at high concentrations on the leaves of many plants. Many strains of this bacterium produce an ice nucleation protein that is apparently located on the surface of the bacteria cell, and the presence of this protein cause the formation of ice at temperature only little below 00 C, inflicting significant frost damage on important crop plants & so facilitating invasion of plant tissues by the bacteria.

(b) Use of nitrogen fixing bacteria to improve crop yields: All animals & plants & most bacteria depend on the availability in their environment of some form of "combined nitrogen" nitrate (No3), or such nitrogen containing organic compounds as amono acids. The huge amount of N2 that exist in the atmosphere are unavailable to the biological world except through the process of nitrogen fixation.

The biological process of nitrogen fixation does not require the consumption of fossil fuels or electricity does not produce environmental pollution. Thus increasing the extent of biological nitrogen fixation has been an important goal for biotechnology. According to one estimate, the biological process fixes times more nitrogen (24 x 10 17 tons/year) than is converted into chemical fertilizers by industrial processes. Thus even a small improvement in exploiting the biological process would have been a global impact.

He biological process of nitrogen fixation is complex & consumes a large number of ATP, molecules, because the enzymes involved in it must overcome the same huge activation energy barriers that faces the chemical synthesis of ammonia. Two enzymes are required: component I (nitrogenase) & component II (nitrogenase reductase). After component Ii is reduced by a strong biological reductant, 16 molecules of ATP are hydrolysed to accomplish the reduction of component I. Reduced component II alone is not capable of overcoming the activation energy barrier. Reduced component I finally reduced N2 to two molecules of NH3. Nitrogen fixation is strongly reductive reaction, and the enzymes involved are usually irreversibly inactivated when they are exposed to oxygen.

The species of closiridium & klebsilla fix nitrogen only under angerobic conditions. Other free living baueria, however, can fix nitrogen even under aerobic conditions. The cyanobaueria, which carry out oxygen-evoluing photosynthesis, perform nitrogen fixation only in specialized cells called heterocysts, which do not produce oxygen. Azotobactor consumes oxygen at an extremely high rate, and in doing so apparently protects its nitrogen fixing machinery.

Another group of bacteria fix nitrogen only when they are in a symbiotic nitrogen fixer is Rhizobium, which invades the root tissues of teguminous plants, such as alfalfa, pea, clover and soyabean and lives in intracellular vacuoles. The vacuoles become filled with an oxygen-binding protein, leghomoglobin produced by the plants. This creates an environment low in free oxygen where Rhizobium is able to carry out nitrogen fixation.

Production of Transgenic Plants

To produce transgenic plants, the crucial steps are the introduction of the cloned genetic material into the plant cell nuclear & the facilitated integration of the cloned gene into the plant chromosome. It is interesting that the best method for doing so uses a system that already exists in nature, the system by which the plant pathogenic bacterium Agro-bacterium tumefaciens injects a portion of its plasmid DNA into plants & inserts it into the plant become.

The Agro-bacterium system is an excellent system for introducing foreign genes into the chromosomes of plants. The transfer into intact plant cells occur at high frequency, the T-DNA is usually integrated into the plant chromosomes at high frequencies without undergoing structural alterations, and the cells that have received the T-DNA can be selected easily by using antibiotic resistance such as the neomycin resistance maricer.

Finally the transgenic plants produced in this manner are quiet stable for at least several generations. Consequently almost all existing transgenic plants with potentially desirable traits have been obtained by this method.

Herbicide Resistance Plants

The advantages of making crop plants resistant to herbicides are obvious. Although there is a fear that such resistant may eventually increase the use of herbiude chemicals, there are also reasons to except that these transgenic plants will promote the use of sager, more biodegradable herbicides, perhaps in smaller amounts one example involves the herbicide glyphosate which inhibits 5-enol-pyruvylshikimate 3-phophate synthase-an enzyme involved in the biosynthesis of aromatic amino acids. This enzyme has been purified from crop plants & sequenced and DNA Probes corresponding to its amino acid sequence have been synthesized. These probes were used to isolate CDNA for the enzyme from the CDNA library of a plant cell line known to overproduce 5-enol-pyruvylshikimate 3-phophate synthase.

The CDNA was then cloned behind the strong CaMv promoter, and the promoter gene complex was introduced into plant cells via a disarmed Ti plasmid vector. The transgenic plants produce a much higher level of the target enzyme & thus are significantly more resistant to glyphosate. These results are encouraging because glyphosate has very low toxicity to animals is rapidly degraded in soil.

Insect Resistant Plants

Bacitlus thuringiensis is used in the biological control of caterpillars because its sporulating cells contain toxic proteins. The gene for one toxic protein was cloned behind promoters that are effectively expressed in plants & was introduced into plant cells via a Ti plasmid vector, thus producing plants toxic to caterpillar. The major problem with this approach has been the low level of expression of the toxin protein in plants, which is presumably related to the fact that the gene come from a bacterium,. Nevertheless, the method has produced tomato & tobacco plants that proved quiet resistant to caterpillars in field tests.

Cotton however, is often attacked by insects that are more resistant to the B. thuringiensis toxin and the low level of expression of this toxin in transgenic cotton plants provided the plants with little if any protection. Recently, the coding sequence of the toxin was altered extensively to replace codons that are rarely used in plants as well as to preclude the formation of a strong secondary structure in MRNA. When cotton plants were provided with this modified gene, their production of the bacterial toxin increased 100-fold and they showed impressive resistant to common bepidopteran insects than damage unmodified plants.

Some seeds contain high concentrations of protease inhibitors, which are thought to interfere with the digestive process in insects. The cloning of cowpea trypsin inhibitor genes, for example and their transfer to tobacco plants have resulted in good resistance to a wide variety of leaf-eating insects.

Viral Resistant Plants

The method most frequency used for producing virus-resistant plants sprang from the observation that, of ten times, plants infected by nearly avirulent virus are thereafter resistant to super infection by a related, highly virulent one. Thus, deliberate infection with avirulent virus strain has been used to produce protection in crop plants. The method is not totally safe, however, because the avirulent strains many mutate to produce strains that are significantly pathogenic. Most plant viruses are positive strand RNA viruses covered by coat protein sub units. When the virus enters an injured plant cell the replication process begins, starting with the progressive uncoating of the virus from the 5-end of the RNA. Tranagenic plants, whose genomes contain introduced tobacco mosalc virus (TMV) coat protein genes and which continuosly synthesize the coat proteins, show resistance to virus infection. The phenotype of these plants is consistent with the idea that resistance is a result of interference with the uncoating of the virus particle. Although the plant cells are resistant to infect TMV particles, they remain sensitive to the TMV RNA or to partially uncoated TMV particles.

The coat protein gene is usually put behind a strong promoter such as the 35S promoter of CaMV & is introduced into a plant's genome via the Agro-bacterium Ti system. Transgenic plants showing significant resistance to TMV alfalfa mosaic virus, cucumber mosaic virus, tobacco streak virus & tobacco rattle virus have already been produced in this manner. In a recent experiment two virus & potato virus Y, were introduced simultaneously into a commercially important potato cultivar, and one of the resulting trangenic plants provide quiet resistant to both viruses under field test conditions.

4

Biotechnology for the 21st Century: New Horizons

Introduction

Biotechnology has been employed by humans for millennia; traditional applications include production of beer, cheese, and bread. But the recent developments in molecular biology have given biotechnology new meaning, new prominence, and new potential. It is this "new biotechnology" that has captured the attention of scientists, financiers, policymakers, journalists, and the public, even though the revenue it generates is as yet only a fraction of that produced by traditional biotechnology. Through the use of advanced tools, such as genetic engineering, biotechnology is expected to have a dramatic effect on the world economy over the next decade.

A variety of U.S. industries, primarily the agriculture and pharmaceutical sectors, already are profiting from this trend; an estimated $7 billion in sales were generated in 1993 as a result of the new biotechnology. These sales are expected to reach approximately $50 billion in the next decade. The United States leads the world in technological innovation and new company formation in biotechnology. The biotechnology industry is responsible for approximately 100,000 high-skill jobs generated by 1,300 biotechnology firms.

Examples of new products derived from modern biotechnology include tomatoes with an extended shelf life and a drug, tissue plasminogen activator, used to dissolve blood clots during heart attacks. But these innovations offer only a hint of the enormous potential of biotechnology. Many other new products are also in development, including disease-resistant plants, "natural" pesticides, environmental remediation technologies, biodegradable plastics, novel therapeutic agents, chemicals, and enzymes that will reduce the cost and improve the efficiency of industrial processes.

A coordinated Federal research effort can provide the leverage needed to fulfil the broad promise of biotechnology, which may well play as pivotal a role in social and industrial advancement over the next 10 to 20 years as did physics and chemistry in the post-World War II period.

The Federal Role in Biotechnology

The Federal Government has propelled the development of modern biotechnology by supporting research that led to the cloning of the first gene in 1973 and the subsequent development of DNA sequencing technologies. A strong foundation has been laid that places the United States at the forefront of biotechnology research. Critical to continued innovation in all spheres of biotechnology is an ongoing Federal commitment to basic research into fundamental life processes, as well as support for the requisite infrastructure.

To date, the Federal investment in biotechnology has been focused primarily in the health field. That approximately 80 percent of the Federal biotechnology investment supports health-related projects and basic research that is broadly applicable but often health oriented. Much of this support comes from the National Institutes of Health (part of the Department of Health and Human Services), as indicated.

The results of health-related biotechnology research are having a profound impact on medicine and health care, providing improved approaches to the diagnosis, treatment, and prevention of disease. While health-related research must remain a national priority, researchers are poised to build on the foundation in basic science to bring the power of biotechnology to bear in other fields. Only 12 percent of the Federal investment in biotechnology research supports applications outside the health arena.

This report presents the overall Federal perspective on biotechnology issues, opportunities, and priorities in areas beyond health that need to be addressed by funding agencies and the research community. The report is the result of a planning effort by the Biotechnology Research Subcommittee (BRS) of the Committee on Fundamental Science of the National Science and Technology Council (NSTC). The BRS is an interagency committee charged with coordinating the Federal investment in biotechnology.

Because biotechnology research is broadly applicable to diverse government missions and goals, it is supported by 13 Federal departments and agencies, all represented on the BRS:

Agency for International Development (USAID)

Department of Agriculture (USDA)

Department of Commerce (DOC)

Department of Defence (DOD)

Department of Energy (DOE)

Department of Health and Human Services (DHHS)

Department of the Interior (DOI)

Department of Justice (DOJ)

Department of State (DOS)

Department of Veteran Affairs (DVA)

Environmental Protection Agency (EPA)

National Aeronautics and Space Administration (NASA)

National Science Foundation (NSF)

This report is aimed at a wide audience, including the U.S. Congress, Executive Branch officials, industry, and academic institutions.

The Federal Government is one of three partners, along with the industrial and academic communities, in the collaborative venture that is biotechnology research and development. The government's role is to sponsor research that will generate new knowledge and to ensure that technological advancement contributes to the overall public good and national welfare. Support of basic research generates innovations, from which previously unanticipated products may arise. The private sector completes the technology development cycle and ensures distribution of beneficial products to consumers.

Report Structure

The resources of Federal agencies, as well as expertise from the private sector, were brought together to identify opportunities in four rapidly developing areas of biotechnology research. Each area is addressed in a separate chapter of this report, which identifies priorities for Federal investment and summarizes current directions and specific research opportunities. (Related activities of each BRS agency are summarized in the Appendix.)

The chapters emphasize multidisciplinary, long-term research, which will be essential in order to enhance the efficacy of existing biotechnologies and develop new ones. The four areas highlighted are agricultural biotechnology, environmental biotechnology, manufacturing/bioprocessing, and marine biotechnology and aquaculture. The concluding chapter of the report outlines infrastructure needs — facilities, human resources, databases, and mechanisms for collaboration and information exchange. A sophisticated infrastructure is required to support all areas of biotechnology research.

Based on the common themes identified in the four areas, the BRS identified three overarching priorities for Federal investment in biotechnology research. Pursuit of these priorities will ensure the development of a strong research base, which in turn will facilitate product commercialization and help maintain U.S. leadership in biotechnology.

Opportunities in Agriculture

Biotechnology offers efficient and cost-effective means to produce an array of novel, value-added products and tools. It has the potential to increase food production, reduce the dependency of agriculture on chemicals, lower the cost of raw materials, and reduce the negative environmental impacts associated with traditional production methods. In addition, the new knowledge gained through basic research into the nature of life and ecosystems at the molecular level can lead to improved farming practices and diagnostic tools for use in agriculture.

Agricultural biotechnology has the potential to produce billions of dollars in revenue per year in the next century. It will play a crucial role in promoting the nation's economic growth, improving environmental quality, and assuring innovative scientific research. Federal support for biotechnology research in this area is essential, in order to build the broad knowledge base needed to commercialize new and improved agricultural products and tools.

The Promise of Agricultural Biotechnology

Due in large part to scientific advances in crop breeding and farming techniques, world food production has doubled since 1960, and productivity from agricultural land and water usage has tripled. Today, the peoples of the world farm an area about the size of South America; without the scientific advances of the past 30 years, farmland equivalent to the entire Western hemisphere would be required.

But a dilemma lies ahead. The world's population is expected to double by the year 2030 to 12 billion, and it is not clear whether current food production is keeping pace with population growth. The question is how best to feed billions of additional people without destroying much of the planet in the process. The disappearance of tropical rain forests, wetlands, and other vital habitats will accelerate unless agriculture somehow becomes more productive and less taxing to the environment.

It seems certain that agricultural biotechnology will play a major role in resolving this dilemma. Biotechnology can be employed to improve the quality of seed grains; increase protein levels in forage crops; and instill in crops resistance to disease, insects, and viruses, as well as tolerance for droughts, floods, and extreme temperatures. In addition, biotechnology can make foods healthier and more nutritious. For example, tomatoes and other fruits and vegetables containing increased levels of certain nutrients, such as vitamins C and E and beta carotene, may help protect against risk of chronic diseases, such as some cancers and heart disease.

Although major genetic improvements have been made in crops, progress in conventional breeding programs has been slow. Moreover, most crops grown in the United States produce less than 50 percent of their genetic potential. These shortfalls in yield are due in large part to the inability of crops to tolerate or adapt to environmental stresses, pests, and disease. For example, some of the world's highest yields of potatoes are in Idaho under irrigation, but in 1993 both quality and yield were reduced severely because of cold, wet weather and widespread frost damage during June. And some of the world's best bread wheats and malting barleys are produced in the north-central states, but in 1993 the disease Fusarium (headblight of wheat and barley) caused an estimated $1 billion in damage.

Major advances also have been made through conventional breeding and selection of livestock. Feed efficiency (1) for poultry and swine has been increased by nearly 50 percent. Milk production per cow has more than doubled since 1955. Diseases such as hog cholera and pests such as screwworm have been eradicated. Yet major diseases of livestock still go uncontrolled and some still go undiagnosed or even unrecognized. These and many other agricultural production hazards have defied traditional solutions such as classical breeding approaches.

New research is needed to address these problems, in order to minimize risks; improve the financial stability of farms, rural communities, and agribusinesses; and assure the long-term competitiveness of U.S. agriculture in the world market. These advances must be accompanied by reduced dependency on pesticides and improved environmental sustainability.

Biotechnology can compress the time frame required to translate fundamental discoveries into applications. With improved technology and knowledge about agricultural organisms, processes, and ecosystems, opportunities will emerge to produce new and improved agricultural products in an environmentally sound manner.

This chapter highlights five broad priorities in agricultural biotechnology research that merit attention by Federal agencies:

- Continue mapping and sequencing of animal/plant/microbial genomes to elucidate gene function and regulation and to facilitate the discovery of new genes as a prelude to gene modification.
- Determine biochemical and genetic control mechanisms of metabolic pathways in animals, plants, and microbes that may lead to products with novel food, pharmaceutical, and industrial uses.
- Extend understanding of the biochemical and molecular basis of growth and development including structural biology of plants and animals.

- Elucidate the molecular basis of interactions of plants and animals with their physical and biological environments, as a basis for improving the organisms' health and wellbeing.
- Enhance food safety assurance methodologies, such as rapid tests for identifying chemical and biological contaminants in food and water.

This chapter identifies, within each priority area, selected areas where additional research is needed to assure a steady supply of high-quality agricultural products at reasonable prices for decades to come. The report also highlights recent scientific advances involving Federally supported research.

Gene Sequencing and Mapping

PRIORITY: Continue mapping and sequencing of animal/plant/microbial genomes to elucidate gene function and regulation, and to facilitate the discovery of new genes as a prelude to gene modification.

To locate desired genetic traits of organisms and effectively use molecular probes to identify DNA sequences, researchers need precise and relatively complete genetic maps. The combination of detailed maps and DNA sequences of genomes can provide a guide to plant and animal breeders for the development of new, efficient, and informed breeding strategies.

The Federally funded Human Genome Project is rapidly advancing gene mapping and sequencing technologies as well as data management techniques for analysing genome data from humans and other organisms. These advances have the potential to make important contributions to the genetic improvement of crops and animals. The agricultural research community is in an ideal position to apply the technical and scientific advances from the Human Genome Project to address the biology of organisms important to agriculture, aquaculture, and forestry.

Plant Genome Research

The U.S. agricultural research community has mounted a considerable effort to understand the structure, function, and regulation of genes in a wide range of crop and forestry species. There may never be resources sufficient to sequence completely the genomes of all agriculturally important plants, but there is a high degree of similarity among species. Therefore, it is cost-effective to select one representative plant for concerted sequencing of its entire genome.

The model experimental plant *Arabidopsis thaliana* offers the best chance for achieving a completely sequenced plant genome in the foreseeable future. *Arabidopsis* has the smallest known genome of any flowering plant, with 100 million base pairs and little repetition in its DNA. Detailed genetic and physical maps have been developed, and understanding of the plant's biology is growing rapidly. A complete understanding of the *Arabidopsis* genome would offer enormous potential for improving agriculture. In addition to studies on *Arabidopsis*, parallel studies will be required at some level with plants used to grow food and fibre.

Knowledge of a single gene can have broad impact. An example is a recent breakthrough in the identification and sequencing of a tomato gene that confers resistance to a bacterial disease. The sequencing of this gene has given scientists the first unambiguous glimpse of the resistance response of a plant to one of its pathogens. This discovery has opened the way to identification of disease resistance genes in other plants as well as created an avenue for investigating the molecular basis for disease resistance.

Genetics of Rhizosphere Microorganisms

Most farms grow only three or four crops, and some grow the same crops year after year in the same fields. These practices promote soil infestations by pathogens that damage or destroy roots. For most crops, soilborne plant pathogens generally are uncontrolled, except by limited crop rotations and some host plant resistance. Recent studies show that some soil microorganisms associated with the roots of crops carry and express genes for defence of roots against pathogens. Little is known about these beneficial microorganisms, which represent an enormous untapped genetic resource for improving crop production and efficiency of fertilizer use. Identification and genetic studies of these microorganisms could enable their direct deployment or use of their genes in the development of disease-resistant crops.

Animal Genome Research

The agricultural research community has been quick to take advantage of findings from the Human Genome Project, such as the identification of the gene sequence coding for protein C. Protein C is needed for proper blood coagulation and can be used to treat blood disorders such as hemophilia. Scientists have inserted the protein C gene into swine and developed animals that produce protein C in their milk. The use of swine instead of human blood donations for protein C production would avert the potential contamination of human medications with human serum viruses. Also, production costs would be low because a small number of lactating sows would be able to meet the medical demand for protein C.

An organized, coordinated national research program is in progress to enhance understanding of the structure, function, and regulation of genes in economically important animal species. To achieve maximum effectiveness and cost efficiency, this effort is being conducted in cooperation with other international genome mapping programs. One important focus is on traits controlled by multiple genes.

Most genetic traits that are economically significant are encoded and controlled by quantitative trait loci. In animals, these sets of genes control characteristics such as leanness of meat, meat tenderness, disease resistance, and, in dairy cattle, milk quality and quantity.

Recent advances offer the possibility of using molecular markers to improve traits controlled by quantitative trait loci. In some cases, quantitative trait loci can be dissected into individual components, which then become amenable to manipulation. More research is needed to translate the advances made with single-gene traits into advances in understanding of quantitative traits.

Metabolic Studies

Priority: Determine biochemical and genetic control mechanisms of metabolic pathways in animals, plants, and microbes that may lead to products with novel food, pharmaceutical, and industrial uses.

In the past, new products were developed by exploiting natural materials. Antibiotics were derived from microbes, spices and perfumes from plants, and pharmaceutical agents from plants and other organisms. Plants also have served as renewable resources for energy, building materials, and industrial materials. Today, the tools of biotechnology offer new ways to exploit these biological resources to the maximum benefit of society. The genetic manipulation of plant biochemical pathways can create new products for food and nonfood uses that will expand markets and increase the value of agricultural commodities.

Metabolic studies, which examine the aggregate of the chemical processes carried out by an organism, are a logical focus of Federal support. The ability to add or delete specific genes and change the level of enzyme activity in transgenic plants has given scientists a new way of manipulating and exploring plant metabolism. The development and study of transgenic potatoes, for example, has made it possible to identify the key enzymatic steps in the starch biosynthetic pathway and has led to development of a new variety of potato that has more starch and less water.

Only a fraction of the estimated 100,000 compounds – including lignin, cellulose, terpenoids, and lipids – produced by biochemical pathways in higher plants have been characterized.

The dearth of knowledge is even more extreme for lower plants, such as algae and mosses. There is a pressing need to understand the intricate mechanisms that control various pathways. Research on metabolic pathways and their biochemical and genetic regulation should be expanded, because genes of use to agriculture are likely to be found in diverse organisms. Plants may be the organisms of choice for production of specialty chemicals efficiently and in large volume.

Cell Wall Biomass Products

The cell walls of higher plants are the Earth's most abundant form of biomass. The structure and strength of wood and its industrial products are determined by the nature of the cell walls. Cell walls are the major component of forage crops; the chemical composition of the walls apparently affects feed digestibility. Cell walls provide the mechanical stabilization of the plant body and influence the ultimate size and shape of the plant. Therefore, scientists must understand the biochemistry and physics of cell walls before they can employ biotechnology to modify the architecture of plants.

While progress has been made in understanding plant cell walls, many key questions remain. For example, not a single major enzyme involved in the synthesis of cell wall polysaccharides has been isolated. Information is needed about the organization of individual components within cell walls. And the regulation of the synthesis and assembly of cell wall materials in response to developmental and environmental signals is poorly understood.

New Uses for Plant Materials

Higher plants are very efficient sources of renewable organic materials. In particular, corn starch, which is both a useful raw material and an important human and animal food, is produced in large quantities at about the same cost per pound as is crude petroleum. Many genes involved in starch biosynthesis have been cloned recently and it should be possible to explore the development of new kinds of starches for use in nonfood applications. The potential use of starch as a co-polymer in biodegradable plastics is being explored.

Plant storage lipids (i.e., oils) can be produced in large quantities, and they represent a chemically versatile form of biomass. Enhancement of oil yield and composition in plants such as palm and rapeseed depends on increased understanding of the factors that regulate lipid biosynthesis. Progress has been made in identifying genes and enzymes involved in the early steps of fatty acid and lipid biosynthesis. To date, the most promising approach to manipulating oil yield and composition involves modifying the vast variety of fatty acids found as lipid constituents in various wild species. This is technically possible only if the basic biochemistry is elucidated.

Novel Products from Plants

Higher plants synthesize an enormous spectrum of chemical constituents. Many have known value as drugs, biomaterials, solvents, flavorings, fragrances, or coloring agents. Through development of transgenic plants, the purity of many plant-derived chemicals can be enhanced and the range of chemicals expanded. One example is the pending development of plants that produce hydroxylated fatty acids. These oils can be used in applications ranging from hydraulic fluids to nylon synthesis. Similarly, the recent development of plants that accumulate biodegradable thermoplastics illustrates the potential for producing completely new compounds in plants.

Many natural chemicals are important in plant defence mechanisms or in abiotic stress responses. Some chemicals confer important horticultural qualities to plants. Thorough studies have been conducted on the biosynthesis of certain classes of these compounds, such as flavonoids, cyanogenic glycosides, and certain alkaloids. Other classes of compounds, such as terpenoids, have received relatively little attention. Terpenoids are important for plant growth and can be used as essential oils, resin acids, and pigments. An efficient mechanism must be developed to obtain specific biochemical information about these classes of compounds, to complement the extensive ongoing efforts to identify them and elucidate their structures. Biotechnology offers promise for accelerating the characterization of these substances as well as facilitating their production for commercial use.

Growth and Development

Priority: Extend understanding of the biochemical and molecular basis of growth and development including structural biology of plants and animals.

Biotechnology offers opportunities to study a wide variety of genetically regulated growth and development processes. Alteration of these processes can help improve product quality and nutritional and economic benefits. The genetic makeup of plants and animals can be altered by either insertion of new, useful genes or removal of unwanted ones. Discoveries made through biotechnological approaches are changing the way plants and animals are grown, boosting their value to growers, processors, and consumers alike.

This section concentrates on the many opportunities in plant research. The numerous potential benefits range from the development of specialty starch polymers for packaging and oils for use as industrial lubricants to modification of post-harvest traits such as ripening and senescence. The recent development of tomatoes with modified ripening characteristics may lead to extended post-harvest shelf life of fruits, vegetables, and flowers. Information about the biological basis of cold tolerance may result in the development of cold-hardy flowers and ornamentals now grown only in subtropical or tropical climates. The engineering of increased salt tolerance in plants may offset losses due to saline soils.

Plant Flowering

Scientists are gaining their first glimpse into the complex mechanisms of flowering. Most plants use environmental cues to initiate flowering so that individuals of the same species flower synchronously, thus maximizing the chances of successful outcrossing. Scientists have discovered molecules in plants called phytochromes that sense dark and light conditions and send signals that regulate plant activities such as sprouting, growth, and flowering.

There is still much to learn about how flowers develop. Investigations of genes controlling floral meristems – the tissue that forms flowers – promise to help explain the genetic mechanisms that manage the establishment and maintenance of this tissue. Ongoing research also is extending understanding of the physiological processes that take place during the transition from vegetative to reproductive development.

Mechanisms of Plant Fertilization

From apples to zucchini, most foods originate from the fertilized flowers of plants. In tomatoes, for example, several genes have been discovered that act within a flower's anthers, where pollen is produced. Promoters (regions of genes that act as on-off switches) also have been identified. Scientists now are pinpointing the overall molecular mechanisms that cue tomato pollen to fertilize the female part of a tomato flower.

Increased understanding of plant fertilization eventually could lead to incremental improvements in the yield and energy efficiency of crop production.

Some plants have genes that render their anthers infertile. This male sterility allows production of higher yielding hybrid seeds, whereby male-sterile plants are fertilized with pollen from specially selected sources. Genetically engineered systems for controlling male fertility are used to produce hybrid seed of important crops such as oilseed rape, corn, and rice. The technology also is applicable to tomato, lettuce, and a wide range of other crop plants. These advances eliminate the need for costly, labour-intensive hand or mechanical removal of anthers in making hybrid crosses and will have a major impact on the billion-dollar hybrid seed industry. In addition, the knowledge gained by studying anther development and fertilization at the molecular and genetic levels will reveal other approaches that can be used to produce novel varieties of hybrid crop plants.

Biological Basis of Plant Form

Conventional breeding often is aimed at achieving certain visual traits. Once the biological basis for these traits is understood, biotechnology could provide the tools to change plant architecture precisely and optimize plant form. For example, changes in leaf form and/or number could maximize photosynthetic capacity and increase production, and changes in root architecture could facilitate and maximize water and mineral capture and uptake. Also, the tools of biotechnology could be exploited to design crops that thrive on soils contaminated with heavy metals, such as cadmium or mercury. Scientists are studying and beginning to redesign the genes responsible for the activity of phytochelatins, natural compounds in plants that bind these metals and detoxify them.

Animal Growth and Development

Research also is needed to explore the genetic basis of animal growth and development. Technologies have been developed for gene cloning, gene transfer, *in vitro* culturing, and sex determination of embryos, and these approaches are being refined for use with animals. Transgenic embryos or offspring have been produced from rabbits, chicken, fish, sheep, swine, and cattle. Genes can be targeted for expression in specific tissues, but the efficiency of gene transfer methods should be improved, and genes important for growth and development must be characterized further.

Environmental Interactions

Priority: Elucidate the molecular basis of interactions of plants and animals with their physical and biological environments, as a basis for improving the organisms' health and wellbeing. Future gains in agricultural productivity and sustainability depend heavily on the use of biotechnology to improve the health and wellbeing of agriculturally important plants and animals. Research on environmental interactions is emerging rapidly as a powerful means of increasing resistance to stress and disease, with consequent economic and environmental benefits. For example, basic research on the genetic control of biochemical responses to water stress in drought-hardy native plants can suggest how crop plants might be modified genetically for more consistent performance in years of drought. Similar studies are needed to improve the tolerance of U.S. crops to heat stress, winter and frost damage, harm caused by salinity, and other physical stresses.

New crop genotypes also must be resistant to an expanding diversity of pests and diseases. Basic research at the organismal and cellular levels, together with classical breeding techniques, already has paid significant dividends to farmers, consumers, and the environment. Today, U.S. wheat, corn, soybean, barley, and sorghum are grown without fungicides because resistant varieties have been bred. However, these and most other U.S. crops still are subject to major damage from insects and diseases caused by soilborne plant pathogens and insect-vectored viruses, for which there has been no useful source of resistance. As a result, U.S. agriculture continues to depend on pesticides. The survival of high-value horticultural crops depends largely on soil fumigants.

Biotechnology also can help reduce the billions of dollars in agricultural losses caused by animal diseases. There are opportunities to boost the immune competency of animal hosts by genetic manipulation, develop biotechnology-derived regulators of immune function, and identify ways to prevent attachment of disease agents to host cells. Factors affecting the disease-producing capacity of a bacterium or virus can be identified, and strategies can be devised to help the host withstand these pathogens. Through the use of techniques such as enzyme-linked immunosorbent assays, polymerase chain reaction (PCR), and monoclonal antibody-based systems, biotechnology can contribute to major advances in diagnostics.

Plant Pests and Diseases

Virologists have transferred virus genes – such as those for production of virus coat proteins – to plants, thereby conferring resistance to those viruses in otherwise susceptible hosts. Some genes also have been shown to limit virus replication and associated damage. Replication depends on the regulatory and metabolic processes of the host; these processes must be understood before scientists can engineer virus-resistant plant varieties. Another possible way to increase plant resistance to viral infections would be to interfere with the spread of the virus, but this approach would require an understanding of the biochemical mechanisms of spread and transmission.

All crop plants are susceptible to bacterial diseases, which have been difficult to control through plant breeding. Bactericides are not a complete solution, either, because bacteria quickly evolve resistance to them.

The bacterial disease fire blight is destructive to pears, apples, quince, and some ornamental plants. This disease has plagued fruit producers for over three centuries, but its mechanism of attack only recently has been revealed using the tools of molecular biology. This knowledge is expected to provide ways to combat this disease.

Nearly all agriculturally important fruit, vegetable, and grain varieties acquire fungal diseases that can cost growers billions of dollars annually. Information is accumulating rapidly on the molecular basis of plant responses to infection by these pathogens. In many cases, the difference between resistance and susceptibility is the rate of response; if rapid, then the plant is resistant. Scientists have found that deliberately disarmed fungal pathogens can trigger the necessary response in susceptible (i.e., slow responding) plants before the virulent pathogen arrives. This approach has opened a new avenue in disease control: the use of "pathogen derived" biocontrol agents to induce resistance in otherwise susceptible plants. This technique is only beginning to emerge as a way to control heretofore unmanageable plant diseases.

Control of Insect Pests Preying on Plants

Insect pests cause major crop damage. Examples include *Phylloxera,* which has caused devastation in European and Californian vineyards, and *Psylla,* which has ravaged the apple and pear industry of the eastern United States. In this century, the United States has imported and released approximately 800 natural enemies of insect pests; about 40 percent are providing some level of biological control. However, this approach to insect pest management has its limits. Further progress will require increased knowledge of interactions between pests and their natural enemies, and, if necessary, appropriate genetic modifications.

The chemicals emitted by plants in response to feeding by insect pests serve as an attractant to the natural enemies of some insect pests. Corn leaves attacked by beet armyworm caterpillars release volatile compounds that attract wasps that are parasites of these armyworms. Moreover, these parasitic wasps can recognize and fly toward the source of volatile compounds produced in response to feeding by their caterpillar hosts; the wasps ignore the odours released from leaves damaged mechanically, such as by mowing. Expanded knowledge of the "information molecules" used by beneficial insects to find their prey on crops could lead to development of crops capable of producing stronger signals, to attract higher populations of beneficial insects. To exploit fully this approach to biological control, more research is needed on the biochemistry and molecular biology of insect pests and their natural enemies.

Due to insects' evolving resistance to pesticides, and the removal of many of these chemicals from the market, there are few effective means of controlling sucking insects, such as whiteflies, aphids, and leafhoppers. These same insects are among the most important vectors (carriers) of viruses, rickettsia-like bacteria, and mycoplasma-like organisms responsible for major plant diseases. New synthetic chemical compounds to be on the market in the next few years primarily target chewing insects. Biological methods to control sucking insects also are advancing at the molecular level. An example is the enzymatic preparation of synthetic pyrethroids.

The salivary glands of aphids, whiteflies, and leafhoppers contain bacterial symbionts that provide essential amino acids to their insect hosts. Only a few of these symbionts have been characterized. Research is needed on the basic biology, molecular biology, and production of these bacterial symbionts. If the bacterial symbionts could be disrupted in some way, either by manipulating the genes of the symbionts or by engineering an antimicrobial agent into a plant, then an innovative control mechanism could be developed.

Beneficial Microorganisms in Plants

Plants selectively aid the growth of specific types of beneficial microorganisms. Some microorganisms, for instance, have been shown to provide growth factors for plants and protect plants against insect attack and infection. Information about the molecular signals and genetic controls of these associations can lead to the use of these microorganisms to improve the performance — and especially the consistency of performance – of important food and fibre crops, and even to enhance the effectiveness of the microbes themselves.

Animal Pests and Diseases

Work is underway to introduce new genes into animals in order to impart resistance to disease and to identify host cell traits that can be used to enhance resistance to disease agents. For instance, a specific receptor on pig intestinal cells is required for *E. coli* to bind to the cell. Strains that do not carry the receptor– or have the receptor deleted through the tools of biotechnology – are less susceptible to the diarrheal disease caused by the bacterium. Other research on enteric rotavirus infections in pigs has identified specific cell membrane components required for attachment of the rotavirus to an intestinal cell. A complementary molecule can be constructed using biotechnological methods to saturate and block this receptor site.

Porcine stress syndrome is an inherited condition in swine that can lead to meat that is of poor quality or unusable. This condition causes significant losses in the swine industry. At the same time, animals susceptible to this genetic disease have highly favourable muscling with low fat content. A candidate gene for this disease has been identified and research now is needed to develop a DNA probe for the trait, as a means to identify susceptible animals and eliminate them from breeding programs. This research will allow the development of swine with high-quality, low-fat meat but without the trait associated with the disease.

Food Safety

Enhance food safety assurance methodologies, such as rapid tests for identifying chemical and biological contaminants in food and water.

Food-and waterborne illnesses caused by microorganisms pathogenic to humans have been estimated to affect more than 80 million Americans and cost the U.S. economy over $40 billion annually. For example, some 9,000 Americans die each year from food-borne illnesses caused by microorganisms such as *E. coli* in meat and *Salmonella* in poultry. At present, virtually all food inspection is visual. Rapid, accurate, non-invasive tools of biotechnology can help improve the detection and control of food-borne human pathogens as well as chemical contaminants. Federal support is needed in this area to assure that these tools are developed in a timely manner.

DNA Probes

DNA probe kits have been developed for *Salmonella, Listeria, E.coli* 0157:H7, and *Staphylococcus aureus.* Compared to traditional culture-plating methods, these new diagnostic kits offer greater precision, shorter turnaround times, and reduced need for highly trained personnel.

DNA diagnostic techniques for Norwalk viruses also are commercially available. Culture techniques have not been successful in detecting these leading causes of gastroenteritis, making the availability of genetically based techniques critical to the detection and identification of the Norwalk viruses.

In addition to detecting food contaminants, DNA probes and other tools of biotechnology can help reduce levels of naturally occurring toxicants in foods. DNA probes can be exploited in research and plant breeding to isolate genes associated with the biosynthesis of major toxicants, facilitate understanding of the genetic regulatory mechanisms for toxicants in plants, and develop lines of plants with reduced levels of toxicants.

Mycotoxins in food are a periodic threat to food safety. DNA probes could help detect the presence of mycotoxin-producing fungi that grow under certain conditions in plant materials such as improperly dried corn and peanuts. DNA probes could also be used to learn more about the sources of fungal contamination in the environment and as a means to develop management strategies under field conditions.

Biosensors

The development of biosensors offers great promise for improving food processing, analysis, and safety assurance. The highly specific actions of biological molecules can be exploited for use in biosensors that can measure the concentration of specific components in complex mixtures. Enzymes, antibodies, and microbial cells can be immobilized on solid surfaces, and the specific reactions they mediate can be detected by various physical and chemical means. Biosensors are commercially available to detect a variety of sugars, alcohols, esters, peptides, amino acids, cell types, and antibiotics. Development of tailor-made membranes capable of separating molecules based on size, electrical charge, or solubility will accelerate biosensor development as well as the exploitation of biomimetic systems.

Miniaturization and mass production of biosensors could increase their availability and decrease their unit cost. Technologies such as microlithography, ultrathin membranes, and molecular self-assembly have the potential to facilitate the miniaturization of molecular and cellular processes for enrichment, detection, and analysis of chemical and microbiological contaminants in food. Advances in the semiconductor industry have made it possible to combine chemical and biological components and integrated circuits in miniaturized systems. Biosensors can be inserted directly into food processing streams to obtain on-line, real-time measurements of important food processing parameters. Miniature biosensors also could be incorporated into food packages to monitor temperature stress, microbial contamination, or remaining shelf life, and to provide a visual indicator to consumers of product state at the time of purchase.

Systematic Study of Food-Borne Microbes

An appropriately designed long-term molecular biology study could characterize and correlate food-associated microbiological isolates in existing culture collections. Food-related isolates from these collections could be used to screen foods for microbiological contamination.

Foods identified in this way then would serve as focal points for regulatory agencies and critical systems aimed at improving control over growing, shipping, processing, distribution, and other steps during which these foods may become contaminated.

Opportunities in Environmental Biotechnology

Biotechnology can be used to assess the wellbeing of ecosystems, transform pollutants into benign substances, generate biodegradable materials from renewable sources, and develop environmentally safe manufacturing and disposal processes. Researchers are just beginning to explore biotechnological approaches to problem solving in many areas of environmental management and quality assurance, such as:

- restoration ecology (which involves reestablishing the nutrient balance and the structure and function of ecosystems);
- diagnostics, epidemiology, and dispersal monitoring related to human disease agents;
- disease, pest, and weed control in agriculture;
- contaminant detection, monitoring, and remediation;
- toxicity screening; and
- conversion of waste to energy.

Environmental biotechnology is not a new field; composting and wastewater treatment technologies are familiar examples of "old" environmental biotechnologies. However, recent developments in molecular biology, ecology, and environmental engineering now offer opportunities to modify organisms so that their basic biological processes are more efficient and can degrade more complex chemicals and higher volumes of waste materials. Notable accomplishments of the "new" environmental biotechnology include the cleanup of water and land areas polluted with petroleum products.

While some success has been achieved, the potential benefits of the new environmental biotechnology are far from fully realized. Advances in this arena are delayed not only by legal and social barriers (which may be formidable) but also by a dearth of basic scientific knowledge about organisms that may be used in biotechnologies and the ecological systems in which these technologies are to be employed. Only new knowledge acquired through Federally supported basic research can provide the foundation for new environmental applications of biotechnology, facilitate the development of these technologies by the commercial sector, and ensure adequate evaluation and safe application of products without blocking innovation with regulatory requirements.

Research in environmental biotechnology has unique international aspects. International cooperation will be needed to help generate new scientific knowledge in this arena, assure U.S. access to the requisite technologies and genetic resources, and establish markets for the resulting U.S. products and processes worldwide. In addition, environmental biotechnology has tremendous potential for use in developing nations seeking low-cost solutions to environmental problems, such as municipal waste disposal, conversion of agricultural wastes to energy sources, and cleanup of polluted areas.

Here, a research program is proposed in one area of environmental biotechnology – bioremediation, an emerging approach to rehabilitating areas fouled by pollutants or otherwise damaged through ecosystem mismanagement.

A novel research framework is described and specific opportunities are identified that offer potential for significant advances. Five research priorities are identified:

- Develop an understanding of the structure of microbial communities and their dynamics in response to normal environmental variation and novel anthropogenic stresses.
- Determine the biochemical mechanisms, including enzymatic pathways, involved in aerobic and particularly anaerobic degradation of pollutants.
- Expand understanding of microbial genetics as a basis for enhancing the capabilities of microorganisms to degrade pollutants.
- As a standard practice, conduct microcosm/mesocosm studies of new bioremediation techniques to determine in a cost-effective manner whether they are likely to work in the field, and establish dedicated sites where long-term field research on bioremediation technologies can be conducted.
- Develop, test, and evaluate innovative biotechnologies, such as biosensors, for monitoring bioremediation *in situ*; models for the biological processes at work in bioremediation; and reliable, uniform methods for assessing the efficacy of bioremediation technologies.

Bioremediation is addressed as one example of an environmental biotechnology. Because the knowledge required for bioremediation is similar to that needed for the development of many other environmental biotechnologies, the research approach described here is likely to have wide application.

Bioremediation: an Overview

The United States has a large number of identified polluted areas, including land, fresh water, and marine sites that, by law, must be cleaned up. Estimates for the cleanup of Federal lands alone may be $450 billion. The extent of contaminated non-Federal agricultural acreage, mining areas, industrial sites, and aquifers and other water bodies is unknown, but the magnitude of the problem is undoubtedly large and clean-up expenses could be astronomical.

It has been estimated that cleanup of both Federal and non-Federal lands could cost $1.7 trillion using conventional approaches, which would produce noxious waste by-products and thereby impose additional clean-up or environmental costs.

Due to its comparatively low cost and generally benign environmental impact, bioremediation offers an attractive alternative and/or supplement to more conventional clean-up technologies.

Bioremediation has been successful at many sites contaminated with petroleum products. However, it is not always the technology of choice because efficacy and the rate of degradation at any particular site cannot be predicted reliably. Improved predictive and process validation capabilities would help stimulate wider use of this technology. Research also could lead to development of biotechnologies to remediate areas contaminated by metals, pesticides, radioactive elements, other toxic materials, and mixed wastes.

These types of studies could be especially productive at this time. Recent developments in biology have provided new tools and approaches for monitoring the environment and engineering organisms with the capacity to degrade environmental pollutants.

These developments have created unprecedented opportunities for significant advances. Indeed, bioremediation is expected to become an industry with annual sales of more than $500 million by the year 2000.

The United States is among several nations developing bioremediation technologies. Maintaining and enhancing the U.S. position in this arena will require continued investment in the generation of new knowledge needed for the development of new technologies. Investment in bioremediation research has the dual benefits of solving important environmental problems while stimulating the growth of the U.S. bioremediation industry.

Proposed Research Approach

Key to the success of the investigations proposed in this chapter is the overall structure of the research program. The complex environmental milieu in which bioremediation and other environmental technologies will be employed demands an holistic research strategy; the traditional, piecemeal approach will not be adequate. A research program designed using an holistic approach would:

- Form a continuum: Bioremediation research generally is conducted at one of three scales: laboratory, pilot scale, or field trial. To help ensure that results achieved at the first two scales can be translated to the field, the research program should be conceived as a continuum, with investigators working at each scale involved throughout the research conceptualization and planning process. The aim is to translate research findings from the laboratory into viable technologies for remediation in the field.
- Be multidisciplinary: The biological and physical complexity of the field environment, where research findings ultimately will be tested and applied, demands a multidisciplinary research team composed of microbiologists, ecologists, engineers, hydrologists, and other specialists.
- Encompass temporal and spatial variation: A major impediment to achieving the full potential of bioremediation is the lack of access to field sites for iterative research on basic mechanisms, technology development, and process verification. Field research sites should be identified that accommodate long-term experimentation on a large spatial scale.
- Address the complexity of "real world" situations: Many polluted sites contain mixtures of wastes of such chemical complexity that a suite of biochemical processes is required for degradation. Waste mixtures may contain sanitary (household) waste plus hazardous organic or inorganic materials. Hazardous waste is defined legally as toxic, corrosive, reactive, or ignitable. "Mixed waste" generated by Federal weapons and research complexes is a combination of radioactive and hazardous waste. Therefore, research focusing on the degradation of mixtures of wastes by groups of organisms is likely to be most realistic and efficacious.

Focus on Microorganisms

Different types of organisms can be bioremediation agents. For example, the use of plants to concentrate pollutants (phytoremediation) is an emerging research area. This chapter, however, examines bioremediation by microorganisms, the most biochemically diverse and least understood group of organisms on Earth.

Microorganisms (primarily bacteria and fungi) are nature's original recyclers. Their capability to transform natural and synthetic chemicals into sources of energy and raw materials for their own growth suggests that expensive chemical or physical remediation processes might be replaced or supplemented with biological processes that are lower in cost and more environmentally benign.

Microorganisms therefore represent a promising, albeit largely untapped resource for new environmental biotechnologies. Research continues to verify the bioremediation potential of microorganisms.

For example, a recent addition to the growing list of bacteria that can sequester or reduce metals is *Geobacter metallireducens*, which removes uranium, a radioactive waste, from drainage waters in mining operations and from contaminated groundwaters. Even dead microbial cells can be useful in bioremediation technologies. These discoveries suggest that further exploration of microbial diversity is likely to lead to the discovery of many more organisms with unique properties useful in bioremediation.

A tiny fraction of the microbial diversity of the Earth has been identified, and an even smaller fraction has been examined for its biodegradation potential. Research aimed at characterizing this diversity likely would lead to the discovery of novel mechanisms for biodegradation of pollutants. Once the biochemical and ecological nature of microbial biotransformations of pollutants is understood, new bioremediation technologies based on these mechanisms will be possible.

Fundamental Research Areas

Expanded research is needed on the basic ecology of microorganisms and interactions among microbial community members. In nature, microorganisms seldom exist or act as single species; instead, they act collectively as consortia. Research examining bioremediation of polychlorinated biphenyls (PCBs) in the Hudson River, for example, revealed that both anaerobic and aerobic bacteria, acting together, were responsible for degrading the pollutant. (Anaerobes grow in the absence of oxygen, while aerobes require it.)

Additional research of this type will provide a framework for understanding how microbial communities respond to various environmental stresses; how to accelerate *in situ* bioremediation by native microbial communities; and whether the introduction of engineered microbes with enhanced bioremediation potential can survive and function within established communities and help remediate the site.

Such studies are likely to provide corollary insights into aspects of microbial biochemistry important for bioremediation as well as the roles of microorganisms in biogeochemical cycling.

Physiology/Biochemistry Research

Priority: Determine the biochemical mechanisms, including enzymatic pathways, involved in aerobic and particularly anaerobic degradation of pollutants.

Much of the information relevant to bioremediation has come from studies of the genetics and physiology of aerobic bacteria. As a result, the best-known biochemical processes related to bioremediation are oxygen dependent. This characteristic limits their effectiveness in many polluted underground and underwater sites with minimal or no oxygen.

Both biological and physical strategies for improving the supply of oxygen in such sites have been proposed, and research on aerobic organisms must be continued. However, long-term success in dealing with a wide array of polluted sites with little or no oxygen will require information that can be obtained only through increased research on the genetics, physiology, and biochemistry of anaerobic organisms.

In nature, biodegradation in sites with little or no oxygen is mediated by anaerobic and microaerophilic microorganisms. Because of the difficulty of isolating and culturing such organisms in the laboratory, their metabolic diversity and their potential use in environmental biotechnology only recently have been appreciated. These technical obstacles are being overcome with improved cultivation methods, new technologies for identification of microorganisms, and new methods for studying their metabolism *in situ*.

New knowledge about anaerobic microorganisms has expanded opportunities for exploiting their metabolic diversity in bioremediation. Since the discovery in 1982 that some anaerobes can dehalogenate carbon compounds, microbes with this capability have been detected in a wide variety of anaerobic environments, such as the Hudson River sediments. However, to exploit anaerobes for bioremediation, more knowledge is needed about the biology of diverse anaerobic microbes, including how they respond to fluctuating oxygen levels.

Because the concentration of oxygen in soil and water environments is often highly variable (in time and space), biodegradation in nature probably is mediated by both aerobic and anaerobic microorganisms.

This phenomenon can be exploited through the development of anaerobic-aerobic technologies for controlled bioremediation. In other words, the environment can be modified either spatially or temporally to allow the development of various types of biodegradative microbial communities, thereby fostering the degradation of pollutants vulnerable to different agents. Anaerobic-aerobic processes can be developed to exploit the full range of microbial metabolic activity for cleanup of environments contaminated with multiple pollutants.

Genetic Research

Priority: Expand understanding of microbial genetics as a basis for enhancing the capabilities of microorganisms to degrade pollutants.

The successful use of microorganisms in bioremediation depends on the development of a basic understanding of the genetics of a broad spectrum of microorganisms, many of them not yet isolated or studied in any detail. Microorganisms adapted to degrade specific pollutants have been found among populations growing naturally at polluted sites. However, the genetic mechanisms underlying specific adaptations are poorly understood.

In the past, researchers have been unable to conduct genetic studies on these hard-to-culture organisms, but recent developments in molecular biology now make it possible to isolate and study genes of almost any organism. Scientists now can analyse genes that govern a wide variety of metabolic processes, including the degradation of environmental pollutants. Such genetic analyses provide information about mechanisms underlying the operation and evolution of degradation pathways.

While the isolation and characterization of novel microorganisms and genes are important in developing bioremediation strategies, knowledge about the general features of the microbial genome also are needed. Bacteria exhibit a high degree of genetic plasticity, or changeability. Analysis of complete microbial genomes will reveal the nature of this plasticity and suggest how genetic engineering can be used to modify organisms to impart the characteristics needed for bioremediation.

New knowledge about the diversity of microorganisms and the organization of their genes also will help explain how environmental factors influence the expression of genes and the regulation of microbial metabolism. Research has revealed that genes expressed when one compound is present can play a role in the metabolism of a second compound.

For example, some bacteria were found to degrade highly toxic trichloroethylene (TCE) when the less toxic compound toluene was present. Mutant organisms also have been isolated that can degrade TCE in the absence of toluene due to genetic changes that cause the TCE-degrading gene to be "turned on" continuously.

However, bacteria capable of TCE degradation may lack other traits – such as tolerance for heavy metals, salts, and acid soils – needed for their use in bioremediation. Genes for these traits could be added through genetic engineering, thereby increasing the degradation efficiency of native microorganisms.

Recombinant microorganisms with expanded degradation capabilities have been developed recently. Researchers in several countries studied a number of microorganisms, each of which degraded a restricted range of pollutants, and characterized the genes involved. Investigators then combined genes from different species into one strain of bacteria that can degrade multiple types of pollutants.(7) Similar applications of modern genetic techniques should make it possible to tailor bacteria for bioremediation of sites contaminated with specific combinations of toxic compounds.

From the Laboratory to the Field: The Research Continuum

Priority: As a standard practice, conduct microcosm/mesocosm studies of new bioremediation techniques to determine in a cost-effective manner whether they are likely to work in the field, and establish dedicated sites where long-term field research on bioremediation technologies can be conducted.

Priority: Develop, test, and evaluate innovative biotechnologies, such as biosensors, for monitoring bioremediation *in situ*; models for the biological processes at work in bioremediation; and reliable, uniform methods for assessing the efficacy of bioremediation technologies.

Microcosm/Mesocosm Studies

Once organisms or groups of organisms with bioremediation potential have been identified through laboratory screening or genetically engineered microcosm/ mesocosm studies – scale-up studies conducted in large bioreactors or on small, protected areas of land or water – can indicate whether the organisms are likely to perform as desired at field scales. This intermediate step permits control of the pollutant concentration, the numbers of degradative agents, and the physiochemical environment.

More specifically, the dissipation of the pollutant can be quantified and the breakdown products measured and identified using labels and tracers that would be too costly to employ in the field. Changes in the degradative capacity and ecological adaptation of the degrading microorganism can be monitored, and environmental conditions that favour bioremediation can be assessed. If non-native organisms are being evaluated, then their ability to survive and compete with the native microbes can be determined. In addition, potential adverse effects of the bioremediation strategies studied can be detected and mitigated.

Many bioremediation applications benefit from testing and scale-up from the laboratory to the mesocosm level. An example is the use of bioreactors for the immobilization of metals from water or the treatment of industrial and municipal waste, before the techniques are used in the field. When degradation is examined under intentionally varied environmental conditions, the results can suggest whether biodegradation is likely to occur in the field at an acceptable rate with or without oxygen and nutrient augmentation. It is useful to employ systems of various sizes, ranging from small laboratory microcosms, such as soil columns and biofilms, to soil lysimeters (large, instrumented *in situ* blocks of soil isolated by physical barriers).

While the results obtained from laboratory bioreactors cannot be extrapolated directly to predict field performance, research at the microcosm/mesocosm scale can reveal genetic and biochemical changes in organisms exposed to the target pollutants.

This approach may save considerable time and resources by suggesting whether a remediation system has any chance to succeed in the field.

Field Studies

The successful development and application of bioremediation technologies depends on field-based research to verify the efficacy of planned approaches under natural conditions. Field research is a complex undertaking, in part because any number of problems can arise. For example:

- an immunological probe may not be effective due to binding of the substrate to the soil or soil organic matter;
- an engineered microorganism may not be competitive with the indigenous population, or it may not survive at certain times of the year or under specific moisture regimes;
- biocontrol and environmental monitoring technologies are unlikely to be effective in all regions, due to differences in plant communities, soils, climate, disease vectors, land management practices, and economic policies;
- geological and soil conditions may preclude delivery of bioremediative microorganisms to polluted sites; or
- supposedly "clean" sites may continue to leach contaminants as chemical and physical conditions change over time.

A further challenge in field evaluations of bioremediation technologies is the need for data collection over long periods of time and at various spatial scales. It can be difficult, if not impossible, to find suitable field sites. A recent report from a workshop sponsored by the American Academy of Microbiology (AAM) concluded that: *A major problem in the development of bioremediation technology is the lack of field sites that are well-characterized with respect to contaminants, geohydrology, and geochemistry; such sites are urgently needed for understanding the natural events that are taking place and also for the transfer of technology developed in the laboratory to field conditions. An integrated interdisciplinary approach is essential for the application and verification of bioremediation, and this can only be achieved under environmental conditions. Although some sites may already exist, their openness and flexibility of use is unlikely to support more than a few efforts in the bioremediation community.*

In response to this conclusion, the AAM held a follow-up workshop to address strategies and mechanisms for field research. The report from the second workshop recommends that field sites have specific characteristics and defines the goals of the interdisciplinary research to be conducted at such sites.

Although some Federal programs are beginning to address the critical need for field experimentation identified by the two AAM reports, there are no long-term, interdisciplinary bioremediation research programs conducted at dedicated field sites. To accelerate the discovery and development of effective bioremediation technologies, several long-term research sites should be established. Coordinated, interdisciplinary research should be conducted to explore the bioremediation-related characteristics of these sites. Then, multidisciplinary groups of researchers should define cooperatively the basic questions to be answered, the scope of the work to be accomplished, and the methods of validation of field results.

Integral to the research conducted at these field sites would be assessment of the efficacy of the biotechnologies employed. As new bioremediation data are collected and verified, new field experiments could be planned and discoveries transferred to the private sector for development and commercialization.

A coordinated program of long-term research at field laboratories could facilitate the development of risk assessment baseline information, including bioavailability data; promote the transfer of all findings to industry and other potential user groups; generate quantitative information on mechanisms of passive remediation as a bioremediation alternative; foster the development of biosensors for process characterization and validation; and, perhaps most importantly, establish a database that could be used to predict conditions under which bioremediation can be achieved.

Efficacy Testing Methods and Data Evaluation

One barrier to the acceptance and use of bioremediation technologies is the lack of established methods for determining efficacy in the field. Evaluation of bioremediation in the field is difficult due to the complexities of the natural environment, the competing abiotic processes, and the vast temporal and spatial scales involved.

The AAM has identified the lack of adequate performance standards as a key factor undermining user confidence in bioremediation. The recent growth of the bioremediation industry has intensified the problem of determining the effectiveness of a large number of competing processes.

The development of standardized evaluation protocols would be facilitated by research comparing various methods for obtaining evidence. Possible measures include numbers and activities of bacteria, changes in inorganic carbon concentration, or the extent of transformation of pollutants into forms that can be biodegraded more easily.

Although innovation might be inhibited by premature setting of standards or overregulation of the fledgling bioremediation industry, methods for standardized efficacy testing and data evaluation should be developed as an integral part of the industry's growth. Ideally, to ensure objectivity, a government agency not directly involved in site cleanup or regulation would help develop standardized testing methods and materials. As efficacy studies are carried out, a large volume of historical data will be generated. The quality and comparability of such data must be evaluated and controlled carefully.

Biosensors: One Promising Assessment Technology

Currently, *in situ* degradation processes cannot be measured or validated directly; researchers must rely on tracers and gas generation to assess bioremediation processes. Increased investment in biosensor research could lead, over the long term, to improved tools for efficacy assessment. The development of biosensors promises to revolutionize the way pollutants are detected and monitored in the environment.

Unlike standard methods, which rely on analytical chemistry in measuring the total concentration of a pollutant, biosensors can detect the fraction available to microorganisms. Biosensors also have the advantage of being nondestructive and located on-line, meaning that samples do not have to be removed and transported to a laboratory for analysis. Biosensors may utilize either whole bacterial cells or specific molecules (e.g., enzymes or biomimetics) as a detection system. Combinations of biosensors in arrays can be exploited to deal with a diversity of toxicants and pollutants.

One type of biosensor involves linking a gene such as the mercury resistance gene (*mer*) or the toluene degradation (*tol*) gene to genes that code for bioluminescence within living bacterial cells. The biosensor cells can signal that extremely low levels of inorganic mercury or toluene are present in contaminated waters and soils by emitting visible light, which can be measured with fibre-optic fluorometers.

A second type of biosensor employs molecular detectors, which consist of enzymes, nucleic acids, antibodies, or other "reporter" molecules attached to synthetic membranes. Antibodies specific for an environmental contaminant can be coupled to changes in fluorescence to increase sensitivity of detection. Fluorescent or enzyme-linked immunoassays have been derived for a variety of contaminants, including pesticides and PCBs. With sustained long-term research, molecular arrays could be constructed on synthetic membranes and other matrices that would allow the simultaneous detection of a range of contaminants in a variety of environmental substrates.

Modelling

Enzyme redesign is an emerging area of research that holds great promise for improving the bioremediation potential of microorganisms. Enzyme models can be created and manipulated by supercomputers to change primary enzyme structure and then predict resulting changes in enzymatic specificity and activity. Changes that would enhance bioremediation activity then can be engineered into bacteria, and those bacteria tested for a desired result in microcosms and mesocosms. In the absence of reasonable parameters for the biological processes involved in bioremediation, models also can be developed to indicate whether physical factors also are at work. Models could be used to distinguish the important variables from those with only minor effects and to extrapolate results for one geographic region to another, based on predicted patterns of interactions between physical and biological factors.

Opportunities in Manufacturing/Bioprocessing Upstream Processing

Priority: Investigate methods to enhance the efficiency and expand the utility of upstream processing technology.

Upstream processing encompasses any technology that leads to the synthesis of a product as well as the fundamental science and engineering needed to understand product formation. Specific areas of opportunity for Federal research include biocatalysis, metabolic engineering, biomass conversion, bioreactor design and cell culturing techniques, and transgenic animals.

Biocatalysis

Nature's catalysts – enzymes – function highly selectively to accelerate the rate of chemical reactions in biological systems. Enzymes also can limit reactivity toward a particular substrate and limit chemical conversion to a single desired product. For many substrates, enzyme-catalyzed reactions exhibit rates that are many orders of magnitude faster than those of uncatalyzed reactions.

In chemical synthesis and other processes of industrial importance, the ability to mimic the selectivity and rate-enhancing properties of enzymes would be of substantial benefit in reducing economic and environmental costs. In some cases, this approach might provide a commercial means of synthesizing a material where only much less desirable methods were available previously.

At present, about 50 enzymes are used in industry. Most of these industrial uses involve breaking down large molecules into simpler ones. Examples include the proteases in laundry detergents, and the hydrolases used in starch processing. The commercial benefits of enzymology would be enhanced if technologists could expand capabilities for creating complex molecules from simpler ones or transforming existing chemical structures into more active compounds. Production of sweetness-enhanced corn syrup by the enzyme glucose isomerase is one noteworthy example of how a chemical transformation can be carried out on the industrial scale. Synthesis routes that can distinguish between right-and left-handed chemical structures are being pursued for pharmaceutical, food, and agricultural applications. Many industrial firms are exploring the unique capabilities of enzymes to carry out such chemical transformations.

Federally sponsored research can facilitate the development and application of new enzymes of industrial importance. Many research efforts are under way, and solid progress has been made using protein engineering, chemical modification, and recombinant monoclonal antibody technology. More recently, recombinant DNA technology has been used to generate catalysts that exhibit the useful characteristics of naturally occurring enzymes but have been designed either to catalyze chemical reactions not found in nature or to alter biological reactions to broaden their specificity and expand their potential for practical application.

Progress in this field will accelerate with an improved understanding of relationships among protein structure, function, and energetics. More efficient methods are being developed to obtain protein structure using X-ray crystallography, nuclear magnetic resonance (NMR), and neutron scattering techniques, coupled with robust algorithms to treat the data and display the results using computer graphics.

The use of computer graphics to visualize the dynamics of proteins also will enhance understanding of the mechanisms of enzyme action. Progress in obtaining structural data in real time ultimately will help modelers develop algorithms to predict the dynamics of enzymes. Research also should be supported on the thermodynamics and kinetics of prototypical enzyme-catalyzed reactions of interest to the bulk and commodity chemical markets; examples include the reactions involved in corn processing and the action of laundry detergents.

Metabolic Engineering

Metabolic engineering is the use of recombinant DNA technology to enhance the activities of a cell by manipulating its metabolic pathways. To exploit this approach, scientists must develop an improved understanding of cell metabolism. Experimental and mathematical techniques are needed for quantifying the effects of altered uses of raw materials by cells.

Techniques such as electrophoresis can be used to distinguish and quantify cellular proteins produced in response to altered metabolism. In addition, non-invasive techniques such as near-infrared spectroscopy and NMR can quantify intracellular concentrations of raw materials. Also of interest is research on thermodynamics and kinetics of prototypical microbial and microalgal systems.

The first applications of metabolic engineering have incorporated genes for pathway enzymes and transport proteins. In this manner, for example, the bacterium *E. coli* has been induced to produce ethanol. But amplified expression of non-native proteins can affect host cell metabolism — sometimes interfering with the engineering objective — because raw material resources are reallocated. The effects may include a slowing of cell growth and possibly cell death. Often, the expression of small amounts of non-native proteins increases the level of stress proteins. This problem is ubiquitous, occurring in simple organisms such as *E. coli* and in complex cells, such as yeast cells, human cells, the chinese hamster ovary cell (used commercially for protein synthesis), and an insect cell line that is a possible source of proteins for making AIDS vaccines. Federal support for research in this area can promote efforts to overcome or mitigate these undesirable cellular responses.

Biomass Conversion

Biomass includes organic polymeric material – such as lignin, starches, celluloses, and oils – produced by biological processes. Plants and algae are the main sources of biomass, generating many billions of tons annually through photosynthesis. Other sources of biomass derived from human activities include food processing wastes, waste paper, and municipal solid wastes.

The Federal Government should expand its support for research that will facilitate development of commercial products from this abundant raw material. Biomass is a potential source of energy resources such as methane and ethanol, commodity chemicals, animal feed, and specialty products (e.g., flavors, fragrances, pigments). Biomass is an attractive alternative to petroleum-based sources because it not only is a low-cost substrate for production, but also is environmentally friendly; it is renewable and may reduce emissions during processing and improve the biodegradability of end products.

Biotechnology is especially relevant to the commercial use of biomass because it employs natural processes uniquely suited for creating specific products. An example of such a process is fermentation, which converts glucose sugars derived from biomass into commodity chemicals. The components of plant biomass, and products that can be derived from them, include

- cellulose (a mostly crystalline polymer of glucose molecules), which can be hydrolyzed (i.e., broken down) to form glucose, which in turn can be bioconverted into a variety of products, such as ethanol;
- hemicellulose (another heterogeneous, amorphous sugar polymer), which can be hydrolyzed to make sugars such as xylose and glucose;
- lignin (a complex aromatic polymer), which can serve as a chemical feedstock or be burned to provide heat and electricity;
- starch, used by the corn wet milling industry and fuel alcohol industry as the basis for large, commercial-scale bioprocesses; and
- oils. Production of diesel fuel from oil seeds is probably the most well known industrial application of a "renewable" oil; the hydroxy fatty acids from oil seeds can be used in a wide range of products, including cosmetics, waxes, nylons, plastics, and coatings.

Fuels and chemicals can be produced not only from oil seed crops (e.g., soybeans, canola, and crambe), but also from unicellular algae and waste fats from animals. The substantial fractions of oils in these resources can be transformed through a chemical process that changes the large molecules in fats and oils into smaller molecules.

Bioreactor Design and Cell Culturing Techniques

If the great variety of potentially commercial bioprocesses are to be developed and applied, then bioreactors must be designed in which the environment can be controlled precisely to maximize process efficiency. Federal support for research in this area can improve the efficiency of bioprocessing and thereby improve prospects for its wide application.

The design of a bioreactor requires a basic understanding of both chemical reactor design and cell biology. First, designers must understand the effects of reaction rates and stoichiometry, mass transfer, heat transfer, and turbulence and mixing on product distribution, reactor productivity and size, and operational characteristics. These phenomena need to be expressed in accurate but tractable models that can be used for design and optimization calculations.

In order to develop effective cell culturing techniques, designers also must have basic knowledge of cellular functions and protein chemistry. They should understand the molecular, genetic, and metabolic processes involved in the growth of cells and the expression of cellular products; and structure/function relationships in the use of proteins for biochemical conversions.

Transgenic Animals

Transgenic animals can be employed either as biofactories for the production of commercial products or as living models for the study of human diseases and evaluation of pharmaceuticals. The use of transgenic animals for these purposes can be more economical or, in the case of human disease and drug models, more realistic than are conventional alternatives. Federally supported research can improve the efficiency of the processes involved and identify new opportunities.

Transgenic animals are being developed for a wide variety of applications. Transgenic cows are used for production of lactoferrin, a naturally occurring milk protein used in products such as baby formula. Several companies plan to market pharmaceuticals from the milk of sheep and goats and the blood of pigs. Anti-clotting factors used to treat cardiovascular diseases have been produced in rabbits, goats, and mice. Research also is under way to develop transgenic animals – mice in particular – that carry genes associated with human diseases. Such animals might enable researchers to identify specific disease genes or produce human antibodies.

Downstream Processing

Priority: Develop capabilities to recover and purify products from dilute bioprocess streams and develop predictive models to facilitate the design of downstream separations.

Downstream processing includes the cost-effective separation and purification of bioproducts, as well as biorefining. Often, downstream processing is the most expensive phase of producing a substance of biological origin, especially for products with stringent regulatory requirements. A key reason for this high cost is the complex and dilute nature of the aqueous solutions in which bioproducts generally are produced; an inverse relationship has been demonstrated between the price of biological products and the strength of the concentrations from which they must be isolated. This is the case for bulk and commodity chemicals, including ethanol from biological sources.

The high cost of downstream processing means there is significant potential for savings from improved processes. That factor, coupled with the broad applicability of downstream processing research, makes this area a prime target for Federal investment. For regulated bioproducts, such as pharmaceuticals, removal of trace impurities is the expensive step in the purification.

Separation and Purification

The separation and purification of materials produced in a bioreactor is a critical part of a manufacturing operation. The biological products involved range from high-value-added substances used as pharmaceutical agents (e.g., insulin) to lower-cost products including commodity chemicals (e.g., ethanol).

High-value bioproducts are usually fragile molecules, such as proteins or peptides, that require highly specialized and mild processing conditions and may need to be separated from a complex mixture of molecules, including cell debris. This combination of factors makes separation difficult. At present, most separation schemes are scaled-up laboratory procedures; research is needed to improve their performance.

Biological methods hold promise for improving separations. For example, a cell could be modified genetically to suppress production of undesirable by-products and alter the cell wall so it would be permeable only to the desired product. It is also possible to genetically engineer a product so that it passes through the cell wall, has an affinity for certain separation matrices, or is joined to another molecule with desirable separation characteristics. In addition to improving separations, biological approaches also reduce energy expenses, a major part of the cost of separating biological products.

Biological processes one day may offer economical alternatives to current, petrochemically based methods for manufacturing organic acids and alcohols.

However, before these bioprocesses can become commercially viable, nontraditional, lower-cost separation methods need to be developed. Research is under way to develop extracting solvents, resins (separation media), and sorbents that are more selective and have a higher capacity than do current materials. Reversible extraction systems are needed that respond to changes in temperature, pressure, or acidity. Combinations of conventional separation methods and biological methods are being explored to reduce product inhibition, which often occurs in fermentations that produce alcohols and solvents.

In addition, mathematical models of separation steps need to be developed to help reconcile regulatory requirements with basic process conditions during early stages of process development, and to meet the demands of a competitive business environment. Modelling enables scale-up considerations to be estimated very quickly, a capability needed in order to commercialize a bioprocess in an industry where being the "first to market" is a critical element of success.

Biorefining

Biorefining involves the use of microbes in mineral processing systems. Biorefining is environmentally friendly and in some cases enables the recovery of minerals and use of resources that otherwise would not be possible. For example, a significant amount of copper is biorefined from slag heaps. Research now under way or planned addresses the use of microorganisms to bioleach oxide and sulfide ores, and to concentrate metals.

Current research on bioleaching of oxide and sulfide ores addresses treatment of manganese, nickel, cobalt, and precious metal ores. The objective is to identify metabolic pathways in the microorganisms responsible for metal solubilization, and to improve their survival rates and stability. Increased understanding of metabolic pathways will open the door to the manipulation of parameters such as kinetics and metal selectivity, with the aim of enhancing mineral recovery.

An intriguing potential use of biotechnology is *in situ* bioleaching of ore deposits or waste piles. Research is needed to develop a mechanistic understanding of the bioleaching process and to identify environmental and process factors affecting biosystem performance. In addition, bioprocess monitoring schemes and cell-free leaching systems should be developed. Cell-free or cell component systems, in which cellular components rather than live organisms are used for a specific reaction, eliminate the need to provide an environment conducive to survival of a microorganism. Such approaches also tend to enhance control over the reaction, due to the specificity of the cell component.

Biosorption and metal recovery from dilute aqueous solutions is an emerging field of interest, from both a resource conservation standpoint and an environmental remediation standpoint. Microbes, algae, cell wall material, proteins, and other types of biomass have been investigated for use in this application, but economical methods for selective metal recovery have yet to be developed. The Federal Government should support research in this area. Inexpensive substrates for protein immobilization should be developed, and biomolecules with high metal-binding capacity identified.

Use of microorganisms to remove fine particles from aqueous process streams is also of interest. Laboratory-scale studies have shown that certain microorganisms can cause flocculation of fine mineral suspensions, while others can function as flotation collectors or depressants. Extensive research will be needed before the use of extracellular colloidal molecules (such as polysaccharides and exopolymers released by live biomass) can become a viable alternative to current chemical-based systems.

Process Monitoring and Control

Priority: Develop methods for monitoring and control of commercial bioprocessing including reliable real-time sensors.

Process monitoring and control requires knowledge of the current state of every step in a bioprocess and the ability to control the process for optimal results. The most efficient use of upstream and downstream processing depends on knowledge of the state of the system and on control algorithms that can optimize the process and maintain it. This is another area where Federal support for generic research can have significant payoffs for multiple industries.

Control and optimization is challenging due to several characteristics of bioprocesses, including their high degree of nonlinearity (meaning that nonlinear differential equations are required for mathematical modelling), especially in batch and fed-batch bioreactors; and their potential for instability when they involve high-yield mutant or recombinant organisms. There are technical obstacles as well, including the unavailability of reliable on-line real-time sensors and realistic models that capture the complexities of biological systems by identifying and quantifying rate-controlling steps in reactions.

Optimization and control methods should be robust, adaptive, and suited to nonlinear processes. Important components of these systems include biosensors and bioelectronics/bionetworks. Because an optimal process often is operating at the limits of process variables, process control can be very difficult; consequently, design methods should consider simultaneously both the economics and the controllability of the process. The affordability of some production processes, particularly for lower-value products, depends on the development of highly accurate sensors and models and the capability to reproduce bioreactor operation precisely.

Biosensors

Sensors are invaluable in the design and operation of automated and environmentally benign manufacturing processes, and in detection, monitoring, and control of food additives, food safety factors, and bioremediation technologies. Biosensors are attractive because they can harness the inherent specificity of many biochemical processes to allow for accurate and highly specific detection of organic and inorganic species in real time. Biotechnology can be used to design, construct, probe, and alter sensitive and highly specific biosensors and incorporate them into transducing and reporting systems.

A biosensor encompasses three major components: a biological component, an interface, and a transducing element. The biological component (e.g., enzyme, immunoprotein, nucleic acids, whole cell) interacts specifically with the substance to be detected. The interface component (e.g., polymeric thick or thin film, chemically modified surface) links the biological component with the transducer. The transducer converts the biochemical interaction into a quantifiable electrical or optical signal.

An example of a biosensor is the plasmon sensor, which consists of a monolayer of immunoproteins immobilized on the surface of a thin film of gold. The analyte— whatever compound binds to the protein employed — attaches to the immunoprotein and changes the properties of surface electromagnetic waves that can be excited in the gold film. This change, which signifies the presence of the analyte, can be detected optically.

A vast store of fundamental information is available on the biological components used in biosensors. The key challenge now is to improve the manufacturing process so that reproducible, dependable, low-cost biosensors can be mass produced. Federal support can hasten progress in this area and foster the use of biosensors in a broad range of applications.

There are a number of technical obstacles. Poor adhesion of polymer films to the transducer, and hence poor reliability, is a major problem. The film must be compatible with the biological component and also adhere to the transducer surface. Another problem is that insufficient knowledge of methods and effects of immobilizing biological components in/on polymeric films has led to wide variation in sensor response, complicating the manufacture of commercial biosensors. Non-specific binding must be minimized, because the specificity of analyte detection is degraded when other molecules compete for binding sites.

There are also operational difficulties that preclude the use of biosensors in bioprocessing environments. For example, many sterilization techniques (such as heat) either modify or denature biological molecules. There are also problems with miniaturization, sensitivity, dynamic range, and noise reduction related to specific biosensor fabrication strategies. Finally, there is the question of quality control – determining quickly and nondestructively whether the biosensor is operating with prescribed precision and accuracy.

Bioelectronics and Bionetworks

Bioelectronics is an emerging technology that employs biological molecules instead of inorganic materials in conventional integrated-circuit technology or in applications involving unconventional architectures, such as optical processors. The driving force for this research is the possibility of constructing devices on the molecular level and thereby achieving extremely high densities of data storage sites and nano-sized computers.

Biological systems are capable of storing information on the molecular level and processing information along pathways defined on a molecular level. Although biological processes function more slowly than do conventional solid-state devices, this penalty is more then offset by a huge increase in the density of operating units. Various biomimetic or biologically based materials, such as the protein bacteriorhodopsin, are being evaluated for use in bioelectronics.

In living things, data processing is achieved by arrays of neurons. Although the operation of single neurons is well understood, the operation of biological neural networks remains largely unexplored. Recent achievements in the culturing of a monolayer of neurons on a micro-electrode array promise to provide some insight into the operation of neuron arrays. Although neuron devices – bionetworks – would not operate on a molecular scale, they have the potential to form the basis for new computer architectures, including parallel processors.

Appropriate Federal investments to address key research questions could help make the United States a world leader in this promising new arena. The major technical obstacle to development of practical bioelectronic devices lies in determining how to preserve and control the properties of the active state when the bioactive species is immobilized in a artificial membrane. Research to explore the fundamental process of biological self-assembly, a capability inherent in biological molecules such as lipids, DNA, and proteins, may be useful in resolving some of the organizational problems. The pattern inherent in a bi-lipid layer or an alkane thiol monolayer ("smart" organic compounds) on a metal surface can be used as a template for further organization of biomolecules.

Biomaterials Processing

Priority: Expand the development of novel biomaterials, such as biomimetics and replacement tissues, through new tissue engineering and chemical synthesis methods.

Up to this point, this chapter has focused on generic processing technologies used to make a variety of products. This section addresses the processing of one particularly promising category of products – biomaterials. The tools of biotechnology can be employed to endow materials with properties not achievable using more conventional means. This research area deserves steady Federal support because of the important applications for biomaterials, which include biological substitutes for human tissues.

Biomolecular Materials

Due to their diversity, versatility, and unusual combinations of properties, biomolecular materials offer promise for application in virtually all sectors of the economy, including defence, energy, agriculture, health, and environmental technology. Examples of biomolecular materials include silk obtained from spiders, and the ceramics in sea shells.

In addition to their direct use as natural cellular products or modified derivatives, biomolecular materials serve another very important purpose by demonstrating how nature has optimized their physical properties.

Research is needed to clarify how higher-order structure is achieved and how it serves to determine macromolecular function in such a variety of forms. With continued advances in modern biology, molecular genetics, and protein engineering, and with rapid improvements in physicochemical characterization, novel biomolecular materials can be designed and tailor-made to meet specific needs. This, in turn, would expand the possibilities for practical applications of these materials. Standard chemical synthesis has inherent limitations, including the production of compounds with unwanted impurities and by-products.

By contrast, biomolecular synthesis – manufacturing methods based on biological processes – allows precise control, thereby reducing levels of impurities and by-products. For instance, when a cell produces peptide polymers, control of the amino acid sequence is assured by the fidelity of RNA and DNA replicative mechanisms. Thanks to recombinant DNA technology and continuing advances in molecular biology, that same control over uniformity of composition, length, and sequence now is available to the scientist seeking to synthesize and express natural or tailor-made genes for peptide polymers.

By inserting appropriate DNA sequences into microorganisms, algae, and higher plants, investigators have learned how to induce the cells to synthesize polymers (such as polylysine) that they do not ordinarily produce, in quantities sufficient for further study and eventual application. Research is needed to define the classes and specific types of polymers that can be produced in this manner. Federal support could play a key role in advancing research in this area.

Tissue Engineering

Tissue engineering – a term coined in 1987 referring to the development of biological substitutes to restore, maintain, or improve human tissue function– employs the tools of biotechnology and materials science as well as engineering concepts to explore structure-function relationships in mammalian tissues. This emerging technology could provide for substantial savings in health care costs and major improvements in the quality and length of life for patients with tissue loss or organ failure.

Advances in the study of tissue growth and regeneration, at both the cellular and tissue levels, set the stage for practical application of tissue engineering. The culturing of cells in two-dimensional monolayers enabled the study of cellular processes and opened the door to genetic manipulation. Scientists and engineers have begun to view cell culturing as a three-dimensional process, in which external forces on the cells not only may influence cellular products, but also may reawaken the cellular differentiation process. In order to develop living tissue equivalents, it will be important to understand how the cellular environment affects the differentiation process as well as interactions between the engineered tissue and the host.

The first success with differentiated cells came with engineered human skin, now in clinical trials. Scientists also are beginning to explore the potential to grow many tissues in culture. Using stromal cells from human tissue, researchers are developing blood vessels, bone, cartilage, nerve, oral mucosa, bone marrow, liver, and pancreatic cells. Federal support can hasten progress in development of these materials.

Encapsulated cell therapy is an example of a technique under development by industry that employs biomaterials in the treatment of certain serious, chronic diseases. The goal of this approach is to replace cells within the body that have been destroyed by disease in order to augment circulating or local levels of the deficient molecules. Targets for replacement include insulin-producing cells in diabetics and dopamine-secreting cells in patients with Parkinson's disease.

An encapsulated cell implant consists of cells that secrete the desired hormones, enzymes, or neurotransmitters, enclosed within a polymer capsule and implanted into a specific site within the host. Animal studies have shown that the functional activity of secreting cells can be maintained in vivo. The capsule wall is designed to allow passage of small molecules (i.e., glucose, other nutrients, therapeutic molecules) but prevents or retards the passage of large molecules, such as elements of the immune system. Studies suggest that the transplanted cells are protected from destruction and perhaps even recognition by the host's immune system, allowing the use of unmatched or even genetically altered tissue without systemic immunosuppression.

Thus, the use of an encapsulating membrane may overcome two of the difficulties that prevent widespread tissue transplantation into humans: the limited supply of donor human tissue, and the toxic effects of immunosuppressive drugs required to prevent rejection of unencapsulated transplants. This is an important research area where expanded Federal investment could yield significant dividends.

Opportunities in Marine Biotechnology and Aquaculture new and Improved Products from the Seas

Priority: Develop a fundamental understanding of the genetic, nutritional, and environmental factors that control the production of primary and secondary metabolites in marine organisms, as a basis for developing new and improved products.

Priority: Identify bioactive compounds and determine their mechanisms of action and natural function, to provide models for new lines of selectively active materials for application in medicine and the chemical industry.

Most major groups of living organisms are primarily or exclusively marine. Tropical marine environments harbour an especially wide diversity of animals and plants. Many marine organisms are sessile and must employ sophisticated methods to compete for a place to anchor. This characteristic is reflected in part by a metabolism that produces enormously diverse bioactive products – many with no terrestrial counterparts.

Recent research has uncovered unicellular and multicellular microorganisms that are unique to the marine world. Indeed, "... it seems clear that marine bacteria are emerging as a significant chemical resource" and the amounts and characteristics of chemicals they produce.

Federal support for research in this area is essential, both to answer the fundamental questions and to assure that the resulting knowledge is translated into sustainable technologies. While it will be vital to cultivate marine microorganisms that produce novel products, the alternative approach of transferring genes of interest into non-marine microorganisms also should be investigated. For example, the capability to produce a marine polysaccharide – a complex molecule that could be useful as a food additive or a water-resistant adhesive – could be transferred to an easily grown bacterium (e.g., *E. coli* or *Bacillus subtilis*). This approach might be more effective in some cases than would cultivating the marine organism or recreating artificially the long and complicated production pathway for the polysaccharide.

Pharmaceuticals

Many bioactive substances from the marine environment already have been isolated and characterized, several with great promise for the treatment of human diseases. The compound manoalide from a Pacific sponge, for example, has spawned more than 300 chemical analogs, with a significant number of these going on to clinical trials as anti-inflammatory agents. To promote sustainable technology development, the Federal Government should support research leading to new methods for discovery. One approach would be to focus on learning about natural functions, regulation, and production of substances generated by marine organisms, in order to identify potentially important agents. Refined test systems should be developed in order to identify selective agents produced by marine organisms. Rapidly developing assay technology can facilitate exploration of the bioactivity of newly discovered compounds. These assay methods, which employ specific receptors for known physiological agents, require only minute amounts of a test substance and can be automated. (Traditional chemical tests require considerable amounts of test material and laborious measurement processes.) The new tools will make it feasible to test newly discovered compounds rapidly by the hundreds, for a wide spectrum of biological activities.

To date, exploitation of natural agents from the sea has been hindered by problems with limited or sporadic distribution and production. Much more research must be conducted to determine what seasonal factors and life cycle or reproduction states are linked with natural production of an agent. Factors influencing production may include diet, physical and chemical conditions, distribution by phylogenetic affiliation, geographic location, water depth, or associations with symbiotic microorganisms. Knowledge of these factors will be important in developing methods for producing selected metabolites, either from whole organisms or *in vitro* from cell or tissue cultures of plants and animals. Many of these compounds are very large and complex molecules, requiring very elaborate biochemical processes; as a result, it can be difficult to synthesize them or clone all the genes for standard production through fermentation.

Enzymes

Enzymes produced by marine bacteria are important in biotechnology due to their range of unusual properties. Some are salt-resistant, a characteristic that is often advantageous in industrial processes. The extracellular proteases are of particular importance and can be used in detergents and industrial cleaning applications, such as in cleaning reverse-osmosis membranes. *Vibrio* species have been found to produce a variety of extracellular proteases. *Vibrio alginolyticus* produces six proteases, including an unusual detergent-resistant, alkaline serine exoprotease. This marine bacterium also produces collagenase, an enzyme with a variety of industrial and commercial applications, including the dispersion of cells in tissue culture studies. Other research has demonstrated the presence in algae of unique haloperoxidases (enzymes catalyzing the incorporation of halogen into metabolites). These enzymes could become valuable products, because halogenation is an important process in the chemical industry.

Japanese researchers have developed methods to induce a marine alga to produce large amounts of the enzyme superoxide dismutase, which is used in enormous quantities for a range of medical, cosmetic, and food applications. An unusual group of marine microorganisms from which enzymes have been isolated are the hyperthermophilic archaea (previously called archaebacteria), which can grow at temperatures over 100° C and therefore require enzyme systems that are stable at high temperatures. Archaea typically are found in extreme environments, such as hot springs, animal guts, hydrothermal vents, sewage sludge digesters, and hypersaline habitats, including the Great Salt Lake.

Thermostable enzymes offer distinct advantages, many still to be discovered, in research and industrial processes. Thermostable DNA-modifying enzymes, such as polymerases, ligases, and restriction endonucleases, already have important research and industrial applications. Hot springs in Yellowstone National Park provided the first archaeon from which thermostable DNA polymerases were isolated. These novel enzymes (the *Taq*® polymerases) became the basis for the polymerase chain reaction (PCR), a useful technique for studying genetic material. In 1989, thermostable DNA polymerase was designated Molecule of the Year by Science magazine. Comparable enzymes continue to be discovered.

Most enzymes involved in the primary metabolic pathways of thermophilic bacteria and archaea are dramatically more thermostable than are their counterparts living at moderate temperatures. Expanded study of enzymes from thermophilic marine microorganisms will contribute to the understanding of mechanisms of enzyme thermostability and should enable the identification of enzymes suitable for industrial applications as well as modification of enzymes to enhance thermostability.

Biomolecular Materials

Recent research has demonstrated that marine biochemical processes can be exploited to produce new biomaterials. For example, a corporation in Chicago is commercializing a new class of biodegradable polymers modelled on natural substances that form the organic matrices of mollusk shells. Equally exciting are the mechanisms used by marine diatoms, coccolithophorids, mollusks, and other marine invertebrates to generate elaborate mineralized structures on a nanometer scale (less than a billionth of a meter in size). Nanometer-scale structures can have unusual and useful properties. Research that will enhance understanding and allow engineering of the processes for creating these bioceramics promises to revolutionize the manufacturing of medical implants, automotive parts, electronic devices, protective coatings, and other novel products. Biomaterials also hold promise for counteracting biofouling, which long has been recognized as an extensive and costly problem. Bacterial biofilms form slime layers that increase drag on moving ships, interfere with transfer on heat exchangers, block pipelines, and contribute to corrosion on metal surfaces. Bacterial and microalgal colonization of surfaces is accompanied by settlement of invertebrate larvae and algal spores, eventually leading to "hard fouling" and the need for costly cleaning.

The most effective anti-fouling coatings have utilized toxic chemicals, such as copper and organotins. There is an urgent need for nontoxic biofouling control strategies, due to heightened recognition of the impact that toxic coatings can have on the environment. Research is needed on the attachment mechanisms of marine organisms and the natural products they employ to prevent fouling of their own surfaces.

Molecular approaches to characterizing biofilm structure and development offer considerable potential for finding novel biofouling prevention strategies. It is now possible to determine the genes and pathways involved in regulation and synthesis of bacterial adhesive polymers. Considerable progress has been made in understanding the nature and expression of surface polymers produced by microorganisms such as the nitrogen-fixing *Rhizobium* species and the opportunistic pathogen *Pseudomonas aeruginosa*. Similar approaches can been applied to marine biofilm bacteria, to find the genetic determinants of adhesive production and the environmental factors that regulate synthesis.

Molecular biology techniques also can be applied to determine the basis of natural antifouling mechanisms. Many marine plants and animals remain free of attached bacteria, either because they produce repelling compounds or because their surface structure neutralizes bacterial adhesives. A product generated by the seagrass *Zostra marina* (eelgrass), for example, is an effective agent for preventing fouling by bacteria, algal spores, and a variety of hard-fouling barnacles and tube worms. Molecular characterization of natural fouling resistance could provide new strategies for fouling control. Potential applications include prevention of fouling in industrial pipelines or heat exchangers, improved design of trickling filters or aquaculture circulation systems, and control of biofilm infections of medical implants and prosthetic devices.

Biomonitors

Marine organisms can provide the basis for development of biosensors, bio-indicators, and diagnostic devices for medicine, aquaculture, and environmental monitoring. One type of biosensor employs the enzymes responsible for bioluminescence. The *lux* genes, which encode these enzymes, have been cloned from marine bacteria such as *Vibrio fischeri* and transferred successfully to a variety of plants and other bacteria. The *lux* genes typically are inserted into a gene sequence, or operon, that is functional only when stimulated by a defined environmental feature. The enzymes responsible for toluene degradation, for example, are synthesized only in the presence of toluene. When *lux* genes are inserted into a toluene operon, the engineered bacterium glows yellow-green in the presence of toluene. This genetically engineered system "reports" that biodegradation of a specific chemical, in this case toluene, is proceeding.

Another type of biomonitor that holds great promise is the gene probe, which can be used to identify organisms that pose health hazards or may be useful in research.

Specific gene probes can be employed, for example, to detect human pathogens in seafood and recreational waters; fish pathogens in aquaculture systems; microorganisms capable of mediating desired chemical transformations (e.g., toxic chemical degradation, CO2 assimilation, metal reduction); and specific fish stocks in fish migration and recruitment studies.

Biopesticides

Natural marine products have the potential to replace chemical pesticides and other agents used to maximize crop yields and growth. Continued Federal support for R&D in this area is likely to result in useful natural pesticides that would provide greater specificity and fewer harmful side effects than do conventional synthetic agents. Current U.S. expenditures for all pesticides amount to $47 billion annually; by the year 2000, biopesticides from marine and other sources are expected to capture an estimated 10 percent of this market.

An example of a marine biopesticide in use today is Padan*TM*, which was developed from a bait worm's toxin known to ancient Japanese fishermen. This natural pesticide has demonstrated activity against larvae of the rice stem borer, the rice plant skipper, and the citrus leaf miner, among other pests. More recently, scientists in Montana discovered novel compounds in marine algae and marine sponges containing symbiotic microorganisms. These compounds promoted growth and stimulated germination and increased root and coleoptile lengths in test plants.

Several sponge and nudibranch species produce terpenes, a broad class of aromatic compounds used in solvents and perfumes and known to deter feeding by fish. Extracts derived from these same sponge and nudibranch species also demonstrated powerful insecticidal activity against two species, grasshoppers and the tobacco hornworm.

Biomass for Energy Production

Approximately 40 percent of all primary energy production, or photosynthesis, occurs in the seas. In this process, oceanic plants (phytoplankton, seaweeds, seagrasses) take up carbon dioxide (CO_2) and, with light energy from the sun, convert it into organic carbon (primarily sugars) and oxygen. The oceans contain 50 times as much carbon dioxide as does the atmosphere, and it is estimated that primary production incorporates 35 gigatons (1 gigaton = 1 x 1015 grams) of carbon into marine biomass annually. This abundant source of fuel for energy production has not been tapped commercially because it is not competitive with soybean meal and other easily harvested, traditional sources of biomass, and also because, regardless of the source, biomass is not competitive with other types of fuels.

The Federal Government should continue to support research on the use of biotechnology to enhance biomass production and utility. At least three general approaches are being explored.

First, the enzyme that captures CO_2 for photosynthesis – ribulose bisphosphate carboxylase\oxygenase or "RUBISCO" – is relatively inefficient, so supercomputers are being used to verify structural information, and the enzyme is being redesigned to optimize its function. Second, the chemical composition of biomass can be altered to make it more suitable for particular applications. For example, marine microalgae are being genetically engineered to boost their lipid content, with the aim of providing a source of alternative fuels that is more economical than are conventional sources. Third, biotechnology is being used to convert biomass to ethanol and other alternative forms of energy and chemical feedstocks.

New and Improved Processes from the Seas

Priority: Develop bioremediation strategies for application in the world's coastal oceans, where multiple uses – including wastewater disposal, recreation, fishing, and aquaculture – demand prevention and remediation of pollution; and develop bioprocessing strategies for improving sustainable industrial processes.

Bioremediation

Bioremediation shows great promise for addressing problems in marine environments and in aquaculture. These problems include catastrophic spills of oil in harbours and shipping lanes and around oil platforms; movement of toxic chemicals from land, through estuaries, into the coastal oceans; disposal of sewage sludge, bilge waste, and chemical process wastes; reclamation of minerals, such as manganese; and management of aquaculture and seafood processing waste.

The full potential for marine organisms and processes to contribute new waste treatment and site remediation technologies cannot be realized without enhanced understanding of the unique conditions in marine environments. For example, oxidation-reduction (redox) states can fluctuate in coastal and estuarine sediments. The impact of changing redox conditions on biodegradation of environmental contaminants must be understood before waste management and remediation strategies and predictive models can be developed for contaminated sediments.

Bioprocessing

The emerging discipline of bioprocess engineering involves the application of biological science in manufacturing, to produce products such as biopharmaceuticals and natural bioactive agents. Bioprocess engineering requires an understanding of the biological system employed (such as a marine organism), isolation and purification of a product, and translation of the product into a stable, efficacious, and convenient form.

An emerging area of interest is the potential of marine bacteria and fungi to produce unusual chemical structures with no parallels in terrestrial organisms. Small-scale studies have begun to indicate the richness of marine microorganisms as sources for novel lead structures.

Aquaculture

Priority: Use the tools of modern biotechnology to improve the health, reproduction, development, growth, and overall wellbeing of cultivated aquatic organisms; and promote the interdisciplinary development of environmentally sensitive, sustainable systems that will enable significant commercialization of aquaculture.

Aquaculture, which long has been practiced in Asia and is increasingly popular in the United States, Europe, and South America, will benefit tremendously from the use of new molecular tools and processes. With worldwide seafood demand projected to increase 70 percent in the next 35 years, and harvests from capture fisheries stable or declining, aquaculture will have to produce seven times as much seafood as it generates now to supply global demand by 2025. The use of modern biotechnology to intervene in the rearing process and enhance production of aquatic species holds great potential not only to meet this demand, but also to improve U.S. competitiveness in aquaculture.

The U.S. aquaculture industry has grown rapidly in recent years. Farm gate receipts exceeded $800 million in 1992 – a fourfold increase since the early 1980s.

The growth and international competitiveness of the U.S. aquaculture industry will be determined by the size of the resource investment in research and technology development. This investment should be made through a partnership of Federal and state agencies and the private sector. The Federal role is to provide leadership in supporting research to advance knowledge in important research areas and to facilitate the transfer of promising results and technologies to the private sector.

The major research issues in aquaculture are similar to those for other agricultural sectors, but the knowledge base for aquaculture is comparatively meager. Development of this knowledge is a particular challenge due to the diversity of cultured aquatic species and the systems for their production. Federal support for biotechnology research in this area will expand the knowledge base and yield significant dividends.

The application of biotechnology promises significant benefits to both producers and consumers of aquacultural products. The use of genetically enhanced organisms may improve production efficiency through improvements in growth rates, food conversion, disease resistance, and product quality and composition. The application of biotechnology to aquaculture also may help conserve wild species and genetic resources and provide unique models for biomedical research.

Enhancing Reproduction and Early Development

Biotechnology can be applied to enhance reproduction and early development of cultivated aquatic organisms. The resulting benefits could include year-round production of gametes and fry of economically valuable species and creation of new markets for specialized, genetically improved broodstock. Similarly, biotechnology may provide techniques for improving the reproductive success and survival of endangered species, thereby helping to preserve the diversity of life on Earth.

As a first step, research should be directed toward improving basic understanding of environmental, hormonal, biochemical, and genetic control of reproduction. More specifically, scientists must identify and understand the mechanisms of expression of genes involved in reproduction and development, improve technologies for cryo-preserving gametes and embryos, improve delivery systems for administration of natural and synthetic hormones, and enhance understanding of the pharmacokinetics of uptake and release of administered hormones.

Improving Health and Well-Being

Biotechnology offers substantial opportunities to improve the health and wellbeing of cultivated aquatic organisms. More than 50 diseases affect fish and shellfish cultured in the United States, causing losses of tens of millions of dollars annually. Biotechnology not only can improve the survival, growth, vigour, and wellbeing of cultivated stocks, but also can reduce disease transfer between cultivated and wild stocks. New products and market opportunities can be developed related to aquatic animal health and wellbeing. The tools of molecular biology can provide a basic understanding of host immunity, resistance, and susceptibility to diseases and associated pathogens by furnishing information about life cycles and mechanisms of pathogenesis, antibiotic resistance, and disease transmission. Improved technologies must be developed for detecting and diagnosing pathogens and diseases and for enhancing the genetic basis of disease resistance, thereby reducing the need for antibiotics and other drugs.

Potential products resulting from this research include gene therapy techniques; broodstock free of pathogens; safe, effective prophylactic agents, including immune modulators, antigens, and vaccines; safe, effective therapeutic agents; and improved systems for administering prophylactic and therapeutic agents.

Improving Quality and Value

The Federal Government has a responsibility to help ensure the safety and quality of food supplies, and biotechnology can and should be an invaluable tool in carrying out this mandate. Biotechnology can be employed to assess and improve the safety, freshness, colour, flavour, texture, taste, nutritional characteristics, and shelf life of aquacultural food products. In addition, practical technologies can be developed to detect and assay toxins, contaminants, and residues in seafood, and to reduce or eliminate contaminants. There are also opportunities to apply biotechnology in improving seafood processing. Research should be conducted to develop and improve technologies for all these applications.

Conserving Genetic Resources

The preservation and enhancement of biodiversity in natural systems is an important Federal priority. Therefore, the Federal Government should encourage and support programs to maintain and enhance biodiversity in aquatic systems through cultivation and stocking of aquatic species. Biotechnology can be employed in two ways to conserve genetic resources of aquatic species.

First, the tools of biotechnology can be used to identify and characterize important aquatic germplasm, including endangered species. Genomes of aquatic species can be analysed and characterized, and quantitative trait loci identified. Second, biotechnology can be applied to improve understanding of the molecular basis of gene regulation and expression as well as sex determination and thereby improve methods for defining species, stocks, and populations. Approaches include developing marker-assisted selection technologies, improving precision and efficiency of transgenic techniques, and improving technologies for the cryo-preservation of gametes and embryos. Ultimately, stocking certain areas with selected, cultivated species and strains could help maintain biodiversity in natural aquatic ecosystems.

Enhancing Biomedical Models

Aquaculture has important purposes other than food production. Because they often adapt to extreme environments, marine organisms can provide unique models for research on biological and physiological processes. Studies of the developmental, cellular, and molecular aspects of marine organisms as model systems will provide insights into the basis of disease mechanisms and pathogenesis in humans. By contrast, use of mammalian organisms as a basis for the development of some types of human disease models may be neither feasible nor cost effective. Progress in this arena will require that sophisticated molecular biology technologies be adapted to marine organisms, in order to enhance understanding of their biological processes. For example, approaches for gene transfer into eggs have been developed for many terrestrial organisms, but not for most marine species. This technology is needed for analyses of gene regulatory systems and gene expression. In addition, methods need to be developed for culturing tissues from marine organisms. Cultured cell lines will provide opportunities for gene transfer and gene expression studies and enhance the usefulness of marine species as biomedical research models. This is an important research area deserving of Federal support.

Understanding and Conserving The Seas

Priority: Improve understanding of microbial physiology, genetics, biochemistry, and ecology in order to provide model systems for research and production systems for commerce, and to contribute to understanding and conservation of the seas.

Scientists have a powerful new array of sampling devices and measuring instruments that will accelerate greatly the acquisition of knowledge about ocean resources and foster their wise use. These technologies include manned deep-sea submersibles, remotely operated vehicles, geosynchronous satellites, sophisticated acoustic measuring devices, pressure-retaining deep-sea samplers, geographic information systems, real-time flow cytometry, PCR and biomonitoring techniques, computerized databases, and other forms of information exchange and analysis.

These tools should be exploited to accelerate the discovery of unknown marine microorganisms and to expand understanding of known varieties. Federal support for this research is essential, because only then will sufficient information be acquired to assure that practical applications will result. As new life forms and processes become known, and as understanding of them grows, marine biotechnology will make significant contributions to the nation's social and economic wellbeing.

Identifying Organisms and their Niches

Biodiversity

Nowhere in the biosphere is biological diversity greater than in the seas, and the extent of this diversity becomes increasingly evident as scientists investigate new environments. In the 1980s, for example, giant tube worms (*Riftia*) were recovered from areas adjacent to deep-ocean thermal vents, and novel mussels (*Bathymodiolus*) that farm methanotrophic bacteria on their gill tissue were discovered around methane seeps in the Gulf of Mexico.

Most newly described species have been and will continue to be microorganisms, although it is clear that new marine plants and animals also await discovery. Less than 1 percent of the extant bacteria— marine and terrestrial — have been isolated and described. The marine environment represents a particularly fertile source for new bacteria, as evidenced by the recent discovery of unusual "cold water" archaea 100-500 meters deep in the oceans.

These archaea comprise a high percentage of the total bacterial ribosomal RNA present in seawater samples, yet they have not been isolated in pure culture and described. Pursuit of this research may provide new understanding of oceanic processes and, once these archaea are cultivated, perhaps a novel source of products. Countless other marine bacteria also have yet to be cultured; advances in microbial culture technology will enable scientists to isolate, describe, and possibly exploit these organisms.

Viruses are another newly appreciated element of marine biological diversity. Electron microscopy and PCR technology have revealed abundant numbers of viral particles in seawater samples. Specific viruses that infect species of marine phytoplankton have been cultured from coastal as well as open ocean sites. This discovery is exciting for at least two reasons. Viral infection of higher forms of marine life almost certainly affects global ocean processes, such as photosynthesis. Marine viruses also will provide new materials for development of genetic and biotechnological tools that can be used to study and manipulate marine organisms. For example, marine viruses could be used to genetically engineer higher forms of marine life.

Marine Ecology

As a basis for developing new applications for marine products and processes, analysing global climate change, and improving fisheries management, scientists must build a fundamental understanding of marine organisms and their specific adaptations to and interactions with their environment. Recent developments in molecular techniques are rapidly expanding capabilities for research on marine ecology. Three research areas seem to be especially fertile and deserving of Federal support.

As a first step, the new molecular tools should be applied to gain insight into the basic molecular and cellular processes by which marine organisms adapt to extreme environments. Examples of basic research that may lead to commercial applications are gene sequencing projects and the ongoing study of special glycoproteins that inhibit ice crystal growth in the tissues of Antarctic fish. These substances, along with gene sequence information for key marine species, may prove useful in industrial and medical preservation processes.

Second, a thorough understanding of marine ecological systems must be developed in order to specify the "normal" baseline level of function and to monitor and predict potential changes and perturbations of systems due to physical, chemical, or biological impacts. The development of predictive models for analysing potential global climate changes depends on the acquisition of fundamental information on molecular regulatory mechanisms of photosynthesis in the oceans.

Third, research on marine ecology can be conducted to benefit fisheries management. The tools of biotechnology can be used to determine the effect of natural and anthropogenic perturbations on the size of commercial stocks, to delineate fishery stocks and management units of living resources, to determine predator-prey relationships, and to restore habitats essential to robust fisheries.

Defining the Impact of the Seas on the Global Environment

Powerful tools are being developed to elucidate the many biogeochemical cycles that determine the fate of all the life-supporting elements on Earth. Scientists are beginning to understand and manipulate the molecular genetics and biology of esoteric metabolic pathways associated with the carbon, sulfur, phosphorous, iron, and other biogeochemical cycles. For example, based on the hypothesis that iron controls photosynthesis in the oceans, immunological probes were used to show that addition of iron to open ocean water off the Galapagos Islands significantly increased energy production.

Marine biotechnology will be useful in assessing the role of the oceans in affecting climate change and the global carbon cycle. Molecular techniques can facilitate and enhance the measurements of CO_2 concentrations and total CO_2 inventories being developed for global ocean models of carbon cycling. There is compelling evidence that the exchange of dissolved and particulate materials between the continental shelf and its boundaries is a significant factor affecting the flux of CO_2 and biogenic elements within the global ocean. Several Federal agencies plan to collaborate on related research, including marine biotechnology applications.

Infrastructure Needs

Physical Infrastructure Needs

Facilities

Funding for facilities has been very low across all agencies. In fiscal year (FY) 1993, only $53.7 million was invested across the Federal Government. Deterioration of existing facilities and lack of funds for new facilities for biotechnology research threaten to limit progress. The broader problem of the deterioration of academic research facilities has caught the attention of the Administration and the Congress, resulting in a substantially increased appropriation for academic infrastructure in the FY 1995 budget of the National Science Foundation. The total scope of this problem cannot be dealt with adequately by any individual agency. While specialized needs must continue to be supported to the extent possible through existing programs, a coordinated interagency effort to upgrade research facilities is needed. There are a number of urgent needs specific to biotechnology research:

- Synchrotron facilities and neutron facilities for structural biology research need to be upgraded and constructed.
- Researchers need to have access to containment facilities for recombinant DNA research at the P-3 and P-4 safety levels. This need is especially acute in some areas of the country and for academic institutions with relatively small, but nonetheless important, commitments to biotechnology research.
- Plant growth chambers and greenhouses need to be constructed, repaired, and replaced.
- Animal facilities need to be constructed, renovated, and maintained to meet animal care guidelines and regulations emanating from the Animal Welfare Act.
- Large, specialized containment facilities, analogous to greenhouses, for bioremediation research, need to be provided to permit controlled growth of microorganisms as well as contamination studies on scales that mimic those of natural sites. In addition, one or several examples of large contaminated natural environments should be selected as field sites for bioremediation to permit studies on appropriate physical and temporal scales over long time periods.

Instrumentation

Equipment and instrumentation are critical not only to support the research of high-caliber scientists, but also to help attract and train the next generation of biotechnology researchers and technologists. Typically, the equipment and instrumentation required for biotechnology are very sophisticated and therefore expensive to acquire, operate, and maintain.

One cost-effective solution to this problem is development of instrumentation centres available to researchers from government, academia, and all industry sectors. Many different types of instruments are required:

- Specialized chemical hoods, large fermentors, and autoclaves are needed to deal with large volumes of contaminated materials in environmental research.
- New, highly sensitive and portable analytical instruments are needed for *in situ* bioremediation monitoring.
- There are needs for capital-intensive instrumentation, such as X-ray crystallography and nuclear magnetic resonance equipment, computer capabilities (particularly the coming generation of parallel processors), and advanced mass spectrometers for use in manufacturing/bioprocessing research.

- There are needs for specialized equipment for the collection, maintenance, husbandry, and cultivation of marine organisms; specialized fermentation and process-recovery equipment capable of withstanding the corrosive effects of sea water, for the production of marine products; and improved sampling equipment and techniques to replace indiscriminate, mass collections of organisms in the field.
- Equipment and associated instrumentation is needed for culturing of, and bioprocessing with, marine organisms and organisms from extreme environments, including hyperbaric environments.

Funding for the acquisition of state-of-the-art equipment for biotechnology research must be made available. In addition, two related issues faced by academic institutions must be addressed:

- Universities have trouble obtaining funds for equipment maintenance.
- Universities have difficulty taking advantage of Federal instrumentation grants which require matching funds.

Repositories

New and expanded repositories are needed to hold and preserve the many specimens and large amounts of data crucial to biotechnology research. There are many acute needs:

- Germplasm for agricultural research must be collected, characterized, and maintained.
- Genetic stock centres are needed for marine organisms and other nontraditional organisms. The wide range of organisms apart from those traditionally used in laboratory research represents a biotechnology resource that must be developed if the United States is to maintain its strong leadership in biotechnology.
- Repositories must be developed for recombinant DNA reagents generated in genome projects as well as for specialized purposes.
- Repositories are needed for expanded microbial collections and particularly to store mixed cultures that prove promising as bioremediation agents. Current repository practices focus on the storage of pure cultures; the requirements for long-term storage of mixed cultures are poorly understood and must be developed.
- Repositories for marine organisms are needed to make these organisms accessible for biotechnology research. Genetically homogeneous strains of selected organisms at various developmental stages – embryos, larvae, juveniles, and adults – should be produced and made available on a year-round basis.

- International efforts should be made to coordinate culture storage and collection activities so that international standards for technical information can be developed and the exchange of data and cultures maximized.

Databases and Reference Standards

A critical issue in biotechnology is how to manage the massive amounts of information generated and ensure that user communities have access to the data in a useful form.

The solution lies in development of publicly accessible, relational databases, such as those that allow scientists to search across plant, animal, and microbial kingdoms for similarities in DNA sequences. These public databases open the door to discoveries about the fundamental nature of living organisms and ultimately to development of new biotechnology applications. The use of databases as integral tools in biology is in its early stages, and the existing database infrastructure is unlikely to provide an adequate resource base for the future. There are several specific needs:

- New and enhanced databases are needed to store and make available information about genomes, macromolecular structure, biological processes, collections of organisms, engineering design parameters, and other topics of interest to biotechnology researchers. These databases should be interactive, allowing manipulation of the data for process modelling and product design.
- International efforts in database development are essential, and information must be accessible to researchers from all nations. Interagency collaboration in support of computational sciences will help fulfil this need.

Reference standards for quality assurance of biotechnology-based measurements are also important. Reference standards assure accuracy and reliability in biotechnology research and commercialization and allow critical measurements to be compatible among different laboratories. Reliable production of high-quality data reduces the need for duplication of data.

Available reference standards for assigning measured values to proteins, glycoproteins, and DNA fragments are inadequate. Standards for measurement in quality assurance would have a major impact in a broad range of research areas, from genome projects to manufacturing. Quality assessment and comparability of results across different laboratories will be a rapid growth area and major tool for dissemination of information and technology transfer.

Human Resources Needs

Innovation and progress in biotechnology depend on the existence of a supportive, thriving intellectual climate. Broad-based education is needed to provide a continuing supply of well-trained researchers. The challenges of biotechnology require problem solving that crosses traditional disciplinary boundaries. Programs to provide retraining and interdisciplinary training of scientists and technical/vocational staff are essential. Federal agencies should expand training grants that link the disciplines underpinning biotechnology (e.g., biochemistry, microbiology, structural biology, genetics, ecology, engineering) and should support technical/vocational education to provide a work force familiar with the concepts and tools of biotechnology.

The Federal Government also should encourage universities to sustain education and training in disciplinary sciences while recognizing the need for communication and collaboration among disciplines. Experience shows that highly specialized undergraduate and graduate technicians, engineers, and scientists often lack the broad fundamental skills necessary for successful multidisciplinary investigation. This is especially true for specialists in molecular biology.

5

Biotechnology of Vegetables and Fruits

The plant is the most important agent in crop production. Soils, cultivation, fertilizers, irrigation, and other factors, in a sense, are all more or less subsidiary. Soils are modified by cultivation, by adding manure or other fertilizers, by drainage or irrigation, and in other ways with the express purpose of changing the environment so as to stimulate plants to increased productivity. Hence, it is not surprising that from time immemorial extended observations and, later, experiments have been made upon the aerial growth of crops under varying conditions. In fact an almost bewildering array of literature has resulted. But quite the converse is true of the underground parts. The root development of vegetable crops has received relatively little attention, and indeed accurate information is rarely to be found. The roots of plants are the least known, least understood, and least appreciated part of the plant. This is undoubtedly due to the fact that they are effectually hidden from sight. Notwithstanding the extreme difficulty and tediousness of laying the roots bare for study, it is not only remarkable but also extremely unfortunate that such investigations have been so long neglected.

It clearly seems that a thorough understanding of the activities of plants both above ground and below ground and the ways in which these activities are favourably or unfavourably modified by various cultural practices should be basic for scientific crop production. Yet almost countless field experiments, selections, breeding, and testing of varieties, etc. have been carried on in all parts of the world with little or no knowledge as to the behaviour of that very essential portion of the plant, the absorbing system. Similarly, in the study of soils, the greatest attention is given to the problems of the physics, chemistry, and bacteriology of this substratum and rather largely to the cultivated portion of the surface only. The soil, barring the living organisms which it supports, is perhaps the most complex, the most interesting, and the most wonderful thing in nature. Surely it should receive thorough investigation.

But a study of the soil and the way in which its various relations affect yield without a consideration of the essential, intermediary absorbing system is more or less empirical. A complete, scientific understanding of the soils-crops relations cannot be attained until the mechanism by which the soil and the plant are brought into favourable relationships, i.e., the root system, is also understood.

The student of plant production should have a vivid, mental picture of the plant as a whole. It is just as much of a biological unit as is an animal. The animal is visible as an entity and behaves as one. If any part is injured, reactions and disturbance of the whole organism are expected. But in the plant, our mental conception is blurred by the fact that one of the most important structures is underground. Nor is the plant usually treated as an entity; it is often mutilated by pruning, cutting, and injuring the root system, frequently without much regard to the effect upon the remaining portion.

Modifying the Root Environment. – In both field and garden the part of the plant environment that lies beneath the surface of the soil is more under the control of the plant grower than is the part which lies above. He can do relatively little toward changing the composition, temperature, or humidity of the air, or the amount of light. But much may be done by proper cultivation, fertilizing, irrigating, draining, etc. to influence the structure, fertility, aeration, and temperature of the soil. Thus, a thorough understanding of the roots of plants and the ways in which they are affected by the properties of the soil in which they grow is of the utmost practical importance. Before other than an empirical answer can be given to the questions as to what are the best methods of preparing the land for any crop, the type of cultivation to be employed, the best time or method of applying fertilizers, the application and amount of water of irrigation, kind of crop rotations, and many other questions, something must be known of the character and activities of the roots that absorb water and nutrients for the plant and the position that they occupy in the soil.

Adaptation of Roots to Environment. – By more or less profound modifications of their root system, many plants become adapted to different soil environments, others are much less susceptible to change. Among forest trees, for example, the initial or juvenile root system of each species follows a fixed course of development and maintains a characteristic form for a rather definite period of time following germination. The tendency to change when subjected to different external conditions becomes more pronounced as the seedlings become older. But some species thus subjected exhibit much earlier tendencies to change than others and a widely different degree of flexibility is also shown. Hence, certain species, *e.g.*, red maple (*Acer rubrum*), because of the great plasticity of their root systems, are able to survive, at least for a time, in various situations from swamps to dry uplands. The roots of others such as bald cypress (*Taxodium distichum*) are so inflexible that they can grow only under certain favourable conditions and their distribution is thus greatly limited. Great variability also occurs in the rooting habits of fruit trees.

Native herbaceous species usually show great plasticity of root habit, successfully adapting themselves to considerable differences in soil environment. A few seem to be quite fixed in their habit of root growth.

Continued study has shown that many field crops, although governed first of all by the hereditary growth characters of the species or variety, are usually subject to change. Certain varieties are able to adapt their root systems to unfavourable conditions much more readily than others. The wide range in soil and agricultural conditions under which vegetable crops are grown renders them particularly suitable both for the investigations of their root habits and also for a study of the agricultural significance of the differences encountered.

Root Adaptation and Crop Production.—Enough work has been done to show clearly that among garden crops root adaptations frequently occur. These will be pointed out in the following s, although a mere beginning has been made. The field is enormously large and difficult but one rich in possibilities. It seems entirely probable that some of the best-yielding crops are able to outstrip others largely because of their greater efficiency in securing a greater and more constant supply of water and nutrients. On the other hand, the failure of a crop to thrive in a particular soil may be due to a lack of adaptation of its root system to the environment imposed upon it. Both of these conditions are illustrated by the growth of flax in India.

The agriculture in many of these areas is ancient, there have been few innovations, and the soil conditions have had time to impress themselves on the varieties of crops cultivated. A condition of equilibrium between the type of plant and the soil has been obtained, as there has been ample time for the operation of natural selection. When we compare the root system of linseed from the black-soil areas with that of the varieties grown on the Gangetic alluvium, striking differences appear. The roots produced on the black soils (of the Peninsula) are deep, somewhat sparse, and are well adapted to ripen the plant quickly with the minimum of moisture. The type of gearing fits the soil. On the alluvium, where moisture is more abundant and where the aeration of the subsoil is poor, the root system is superficial but at the same time well developed. On the intermediate types of soil... linseed produces a type of root about halfway between that of the black soils and of the alluvium. Once more the root system is found to fit the soil type.

Further when we grow side by side on the alluvium these three classes of linseed, there is little or no adaptation of the roots to the new conditions, but the three types behave very much as they would in their native habitat. The deep, sparse root system of the black soil areas is developed in the alluvium, although it is fatal to the well-being of the crop. When, the experiment is reversed and the types which suit the alluvium are grown on the black soils, there is again little or no adaptation to fit the new conditions. The linseed crop consists of a large number of varieties which differ from one another in all sorts of characters, including the extent and distribution of the roots. The root systems of the varieties are just as characteristic and just as fixed as the differences in the seed and other above ground characters of these plants. A similar state of affairs obtains in other crops like wheat... and the opium poppy, and is probably universal all over India.

Thus the differences between the root systems of varieties have been clearly shown and the economic significance illustrated.

Vegetable growing is an important phase of agriculture and one that is increasing at a rapid rate. In the United States practically every kind of soil is used for growing vegetables. Among the dozens or sometimes hundreds of varieties of the various vegetable crops, some undoubtedly have root systems more suited to a given soil environment than others. If the highest possible production is desired, attention should be given to selecting the variety that is not only climatically adapted but also best fitted to a particular soil, modified as favourably as lies within the power of the grower. It seems certain that a large part of the success in connection with the improvement of varieties consists in better adapting the absorbing system to the soil. As stated by one who has extensively worked on plant selection, breeding, root investigations, and other phases of agronomic science, "The more I learn about cultivated plants the more I am convinced that the future lies in the root-soils relation and in matters which influence it."

An intimate knowledge of the habits of growth of the root systems of vegetable crops will enable the grower to space plants to better advantage. It should also permit him to intercrop or grow in succession such crops or mixtures that the soil volume will have a better distribution of roots and thus permit of methods of more intensive cultivation. Similarly, by means of proper crop rotations and occasionally cultivating very deeply rooting crops, the subsoil may be kept in good condition and the effects of drought mitigated.

Interrelations of Plant, Soil, and Climate. – In considering the importance of root relations in crop production, it should be clearly kept in mind that the plant, the soil, and the climate form a closely interlocking system of which no part should be overlooked or overemphasized. It is now rather generally recognized that climate and vegetation are the most important factors determining the character of the mature soil. 148 "The features assumed by the soil in its development from infancy, through youth, maturity, and old age, vary with the environment, especially with the climate and the vegetation." The effect of both climate and soil on the growth of above ground plant parts has long been known. It has only recently been clearly demonstrated that the environmental factors which affect the root are not only those of the soil immediately about it but also those affecting the shoot which is rightly a part of the complex. Through the shoot the root system is influenced by the aerial environment. The amount of light or the degree of humidity, temperature, etc. and the effect of these upon food manufacture, water loss, and other activities affect root development. In fact there is a rather close correlation between shoot and root development. Whatever affects the above ground growth of plants whether favourably or unfavourably is, in turn, very likely to exert an influence upon root development.

The complex relationship of plant, soil, and climate may be further illustrated in the use of fertilizers. They modify the habit of growth as well as the composition of the plant. For example, phosphates, when applied to soil upon which wheat or certain root crops are grown, promote deeper root penetration. This results in a greater water and nutrient supply for the plant. Earlier development and ripening may be promoted or drought mitigated. Soils thus fertilized will produce crops under an environment perhaps otherwise quite unfavourable. Vegetable production should be studied from the point of view of how roots and shoots of plants grow, and use should be made of the plant itself for indicating the direction of future research.

An adequate discussion of the environment of roots (the soil), how roots are built to perform their work, and root habits in relation to crop production has been so recently given in "Root Development of Field Crops" that further statement seems unnecessary. For a general discussion of the effects of irrigation, drainage, water content, aeration, temperature, nutrients, tillage practice, plant disease, and related phenomena upon root habit and their significance in crop production, the reader is referred to the same volume.

Activities of Roots in Subsoil.—The great extent of the root systems of most vegetable crops and their usual thorough occupancy of the subsoil may at once arouse interest concerning the importance of the deeper soil layers. Experiments have shown that the roots of crops are active in the absorption of both water and nutrients even to the maximum depth of penetration. 174,165 Nutrients, when available, are taken from the deeper soils in considerable quantities, although to a lesser extent than from the soil nearer the surface which the roots occupy first and, consequently, at least in annual crops, where they absorb for the longest time. The deeper portions of the root system are often particularly active as the crop approaches maturity. Nutrients absorbed by them may produce a pronounced effect both upon the quantity and the quality of the crop yield.

Method of Root Study.—In the present studies the direct method of root examination has been employed. It has been used by the writer and his coworkers in the excavation of hundred of root systems during the past 14 years and has proved very satisfactory. By the side of the plants to, be examined, a long trench is dug to a depth of about 5 feet and of convenient width. This affords an open face into which one may dig with hand pick and ice pick and thus uncover and make a careful examination of the entire root system. This apparently simple process, however, requires much practice, not a little patience, and wide experience with soil structure. In every case several plants were examined at each stage of development to insure an adequate idea of the general root habit. As the work of excavation progressed, the trench was deepened, if necessary, so that finally the soil underlying the deepest roots was removed. Frequently, the trenches reached depths of 6 to 11 feet.

Upon excavating the roots, detailed notes and careful measurements were made in the field. After several plants were examined, these notes were studied and any point that remained indefinite was at once clarified by further study. This method leads to a high degree of accuracy. Drawings of the root systems were made in the field on a large drawing sheet with pencil and later retraced with India ink. They were made simultaneously with the excavation of the roots and always by exact measurements. In the drawings the roots are arranged as nearly as possible in a vertical plane; that is, each root is placed in its natural position with reference to the surface of the soil and a vertical line from the base of the plant. In some cases the drawings represent the roots in their natural position in the surface foot of soil. In every case it was sought to illustrate the average condition of root development rather than the extreme. Although the drawings were made on a large scale, the rootlets were often so abundant that it was quite impossible to show the total number. Such drawings, however, carefully executed, represent the extent, position, and minute branching of the root system even more accurately than a photograph, for under the most favourable conditions, especially with extensive root systems, the photograph is always made at the expense of detail, many of the finer branches and root ends being obscured.

Conditions for Growth at Lincoln, Neb

Since roots, like above ground parts of plants, are greatly modified by environment, a brief statement will be given of the conditions under which the crops were grown.

Soils.—The vegetable crops at Lincoln were grown in a fertile, dark-brown, Carrington silt loam. Preceding them white clover had been raised for 2 years. This was followed by a crop of maize. The soil was not only in an excellent condition as regards tilth but also was well supplied with humus. In general, it may be said that the soils of eastern Nebraska have a sufficient supply of all the essential nutrients to insure good crop yields, except that, owing to the systems of cropping, there may be a deficiency in available nitrates. The area was only slightly rolling, the general slope being toward the south. As usual, some differences in soil texture were found in excavating the root systems in different parts of the area. The following description of the soil profile represents an average condition.

The surface 12 to 14 inches was a mellow, very dark silt loam from which the roots were readily removed. At greater depths it intergraded into a dark-coloured, clay subsoil. This became quite sticky when wet and hard when dry. It exhibited a columnar or jointed structure, especially below 18 inches, cracking badly when it shrank upon drying. Roots were removed with much more difficulty from this soil layer. At about 3 feet in depth the subsoil became much lighter yellow in colour and graded rather abruptly into an easily worked, quite mellow, friable soil type approaching loess in many of its physical properties. This extended beyond the depth of greatest root penetration, i.e., over 12 feet. It was characterized by rusty streaks and numerous small calcareous areas and concretions. Earthworm burrows penetrated the soil to depths of 8 feet or more and numerous old-root channels extended almost as deeply. In many places extensive surface cracks reached depths of 4 to 5 feet. They were sometimes 1 inch in width. These crevices had been formed during past years of drought and were filled with rich, black, surface soil which had been washed or blown into them. Cultural operations in a large measure had also undoubtedly contributed to the filling.

Number and Size of Plats.—The crops, in nearly all cases, were grown in triplicate plats so that early, midsummer, and later examinations could be made without disturbing the areas in which the plants were to make further growth. The plats, although somewhat variable in size, were in all cases large enough to permit of normal field development, each group of plants being entirely surrounded by plants of its kind. For example, the plats of lettuce, radishes, and onions were 12 feet long and 10 feet wide, the plants in the central portion of each plat being used for a single root study. In the case of the tomato, cabbage, eggplant, and other crops where each plant occupied considerable space, the plats were proportionately larger so that the plants examined were grown under the ordinary competitive conditions and mutual environment of garden crops.

Tillage.—The field was plowed 8 inches deep early in the spring after all cornstalks, stubble, and dried weeds had been carefully removed. This was followed by repeated harrowing until an excellent, firmly compacted, moist seed bed of good soil structure was formed. On areas not immediately planted, weeds were removed by raking and hoeing. After the seeds were sown or the seedlings transplanted, further cultivation was done very shallowly with rake or hoe so as not to disturb the roots. Cultivation and weeding were repeated, as in ordinary gardening, when needed to prevent the growth of weeds and to keep the surface soil in a good condition of tilth. At no time did cultivation extend to greater depths than 1 inch.

Precipitation.—The rainfall of 19.2 inches during the 6 months of the growing season was 2.3 inches below the normal, with periods of moderate drought occurring in May and again in July.

April with a deficiency of 1.2 inches had three well-distributed showers. In May efficient rains occurred on the eighth and fifteenth, with a monthly deficiency of 3 inches. June was a wet month with a total of 6.6 inches of precipitation, mostly in seven well-distributed rains. This was 2.3 inches above the mean. July had four rains during the latter half of the month and a total precipitation 1.8 inches below the mean. Five well-distributed showers fell in August, an excess of 1.2 over the mean. September had only 0.2 inch above normal, the rains also being well distributed.

As a whole the growing season was typical for eastern Nebraska. The spring was cool, frost occurring until about May 12.

Soil Moisture.-It is a well-established fact that rainfall is only a very general indicator of soil moisture, since many other factors both climatic and edaphic intervene between precipitation and water available for plant growth. Hence, a study of the soil moisture in several of the plats was made from time to time throughout the growing season. Ideally this should have included determinations for each kind of crop at frequent intervals. Owing to the laborious task of excavating roots this was not attempted.

Temperature.—The temperature of the air is an important factor not only directly in promoting metabolic processes but especially in connection with root studies in modifying humidity and affecting transpiration. The amount of water lost from the above ground parts reflects itself in the development and extent of the absorbing organs.

The soil above 6 inches was not only much warmer at times but also subject to greater temperature fluctuations than at greater depths. Thus roots lying close to the surface of the soil were subject to the influence of an environment which was quite different from that affecting those growing more deeply. It seems probable that surface-soil temperatures never became so high as to be injurious to root growth but that the absence of roots or their death in the shallowest soil was due to lack of a sufficient supply of moisture. 15,174 Soil temperatures are also of great importance because of their relation to disease-producing organisms that may greatly limit successful crop production.

Humidity.—The loss of water through transpiration is directly controlled in a large measure by the amount of moisture already in the air. This is also an important factor governing the evaporation of water directly from the surface soil. Consequently, a record of relative humidity was obtained by means of a hygrograph appropriately sheltered with the recording apparatus about 5 inches above the soil surface

Evaporation.—Evaporation was measured by Livingston's standardized, non-absorbing, white, cylindrical atmometers. These were placed in a bare area in the field with the evaporating surface at a height of 2 to 4 inches above the surface of the soil.

Conditions for Growth at Norman, Okla

The greater portion of root studies in Oklahoma was made during 1926. Some work was also done the preceding growing season and the root growth of several crops was followed during the intervening winter. Where root excavations were made in Oklahoma this is so stated in the text.

Soil.—The crops were grown in a fine sandy loam soil that had been heavily manured for a number of years and used for growing vegetables. It was in excellent condition as regards tilth and amount of humus. The area was level and the soil quite uniform throughout. The surface foot had a reddish-brown colour, was quite sandy in texture but rather compact, and contained enough clay to exhibit a tendency to bake when it dried after heavy rains. In the second foot it was a light-chocolate brown and contained more clay. When dry, it became quite hard, especially at a depth of 18 to 24 inches.

The greater sand and smaller clay content of the soil prevented it from cracking and fissuring as did the soil at Lincoln. This, as will be shown, had a marked effect upon general root habit and branching. Roots penetrated the 18- to 24-inch soil layer with some difficulty, especially when it was rather dry. This was shown by their tortuous courses. The third foot consisted of a reddish-yellow sandy clay. It contained numerous, small, black concretions of iron 1 to 2 millimetres in diameter. After long periods of drought it became very hard. The next 18 inches were much sandier and had more and larger concretions but showed no change in colour. At greater depths the soil became still sandier, was mottled with reddish-brown spots, and contained many concretions. At 7 to 8 feet it was very compact.

Method of Planting and Tillage.—The field was plowed to a depth of 8 inches late in March. Repeated disking and harrowing resulted in a good, compact seed bed. All of the crops were planted by hand, each in four to eight rows, 3.5 feet apart in order to permit of cultivation with a horse-drawn harrow. Shallow cultivation was given after each rain to prevent the formation of a surface crust and also to kill weed seedlings. A mulching fork, supplemented by a hoe when necessary, was used near the plants. In this manner a surface mulch 1 to 1.5 inches deep was maintained but in no case was a cultivation deeper. A good supply of moisture was present just below the mulch during almost the entire growing season.

Precipitation.—The rainfall of 11.5 inches during the period when the crops were grown (April to July, inclusive) was 3.6 inches below the mean.

April with 1.9 inches had a deficiency of 1.3 inches and no efficient rainfall after the tenth. May had 2.3 inches which was somewhat (3.1 inches) below the normal amount. No efficient rain fell between May 8 and 30. During June precipitation was 1.6 inches less than the mean. A precipitation of 2.2 inches, however, was fairly well distributed throughout the month. July had 5.1 inches rainfall, 2.4 inches above the mean, little moisture falling after the thirteenth.

Soil Moisture.—Notwithstanding the rather limited rainfall, a good supply of water was present throughout the entire growing season. The fine sandy loam soil of the level field, loose from the cultivation of the previous season, had been thoroughly wet by the fall and winter rains. Water had entered the deeper subsoil and moistened it beyond the depth of greatest root penetration. Most of this water was retained in the spring and supplemented by rains in March. Thus, although the summer rainfall was light, a good supply of moisture was present throughout the entire growing season.

Other Factors.—Another factor favourable to moisture conservation was a very cool, late spring. Warm weather did not begin until the second week in April. The surface soil warmed rapidly, there was enough moisture to promote prompt germination and vigorous growth, and the crops made excellent progress even under a light precipitation. The usual drought and high temperatures of July were replaced by a relatively cool, moist midsummer during which the plants made a continuous growth. The average day temperatures (6 a.m. to 6 p.m., inclusive) and the average daily temperatures (24 hours), obtained by means of a thermograph placed in the field.

Conditions for Growth during the Fall and Winter.—The root development of certain crops was studied during the preceding fall and winter.' Following a dry, late summer in 1925, September was warm and had an abundance of moisture (5.4 inches).

During this period many long-lived vegetable crops showed a vigorous, renewed growth. Growth was continued, especially by the hardier plants, during an unusually cool October. Owing to favourable soil moisture and rather uniformly moderate temperatures during November and December, growth in several crops slowly continued. Following the cessation of growth in January, renewal of the development of both roots and tops was resumed during February which had a temperature of 8.2°F. above the mean.

Introduction Organic Vegetable Growing

Soil & Site

Soil testing: Soil texture and structure. The hand test will tell you which of the soil types you have, Sandy Loam, Medium loam, Heavy Loam. Although Soil Texture cannot be easily changed, knowing the make up of your soil will give indications on how to make the best use of what you have with cultivations and fertility management. In general heavy soils will hold onto nutrients better but will tend to water log in wet seasons. Lighter soils will be hungry well drained and will heat up quickly in spring. Both extremes will benefit from the addition of humus

*Soil nutrients:,*the test for NPK will inform decisions on what methods of fertility management are required and how best to plan the rotation.

pH and lime Test for acidity and recommend lime use. pH has a big influence on what nutrients are made available to the plant and on other soil fauna which are part of a healthy soil. In general terms alkaline soils are better for most vegetable crops whereas acid conditions give opportunities for ornamental plants. Rate of lime application can be calculated using the pH test results.

Site: Choice of site for vegetable patch usually limited,where possible avoid shade, water-logging and overhanging trees and tree roots. In general terms anything that will interfere with access to sun rain and nutrients.

Rotation and Planning

Rotation: Using a rotation of crops means that the same crop is not grown on a plot of land for two years running. Crops are divided into groups of related plants or crops requiring the same conditions and are rotated around pre determined plots within the growing area, so providing the opportunity to apply manures, lime and pest control measures according to the crops requirements. For example by keeping brassicas and potatoes on separate plots allows lime to be applied to brassicas to discourage club root while avoiding lime on potatoes which may encourage scab. The movement of crops around the growing area limits the spread of some soil borne diseases although this is less effective on smaller plots. The rotation can be changed according to need. A traditional rotation would be 4 or 5 years say: Alliums followed by Brassica, followed by Roots followed by Legumes. If Green Manures are used extra years will be added to the rotation. The rotation will depend on what quantities of each crop are to be grown and whether a green manure crop is to be grown or if fertility is to be added using FYM or compost.

Alliums: Have need the highest nutrition due to primitive roots also respond to high nutrient levels. Must have high light levels. It is advisable to plant Alliums immediately after green manure or FYM if used

Brassica & Lime

Brassicas respond well to lime and as it helps prevent the establishment of club root. Lime can therefore be applied if needed before planting. Brassicas can follow Alliums which have had an input from green manure or compost so there is residual fertility from the green manure and any lime applied will help to stimulate microbial activity so liberating more nutrients.

Legumes

Legumes respond well to alkali soils and need reasonable nutrient levels to establish, thereafter they produce some nitrogen from their nitrogen fixing root nodules. Legumes can be flexible where they fit in a rotation as they have less soil associated problems.

Beets and Salads

These can fit with legumes or alliums, they respond well to high nutrient levels and do not have many soil borne problems.

Roots: Potatoes, parsnips and carrots, used late in the rotation so FYM and lime are broken down in previous plantings, so preventing forking of carrot or parsnip roots. Limc avoided for the roots part of the rotation as high lime content in the soil can increase incidence of common scab in potatoes. Compost or FYM can be added for potatoes but should be avoided for carrots and parsnips. The quantity of potatoes grown will determine if they should have a rotation of their own or if they are included with roots or possiblly legumes.

Green Manure: Green manures will replace nutrients used by crop plants if legumes such as sweet or red clovers are used. They can be used to prevent nutrient loss through leaching when non legumes are used notably over winter. They also provide humus and encourage soil fauna activity.

If Green manures are to be used exclusively to replace nutrients and humus they will need to be grown ay least one in three years or more depending on the natural fertility of the soil and density of planting.

Compost

Making a compost heap: Garden style, square up to 2m square by about 2m high, this size ensures that oxygen can reach the inside of the heap while still being big enough to retain the heat needed for efficient working.

There are four factors in making a successful compost heap;

1. Oxygen, make the compost no more than 2m wide so air can pass into the centre of the heap, also the heap should not be over compressed so to prevent the access of air to the centre.
2. Water, necessary for the bacteria to develop and to allow the heap to heat up. When high nitrogen is added to a heap the composting process can be so rapid that it uses up all the available water and stalls the composting process. In winter however the heap must be protected from too much water as this will push the oxygen out causing anaerobic conditions which will also stall the process. Water may have to be added to the heap in hot dry conditions.
3. eat, this will be generated by the composting process but the design of the heap will help to retain it and kep the process going. The heap should be compacted so each layer is in contact with the next but so squashed as it removes all the air. Avoid water logging and drying out.

4. Carbon to nitrogen ratio: The nitrogen (leafy material) speeds up the process while Carbon (wood/straw lignified material) provides the material to turn in humus and slows the process. With too much N the compost will heat up quickly then stall. With too much C the compost will be cold and slow to start.

In the case of a stalled heap it can be restarted by turning the heap adding water, lignified (woody) material, leafy material as needed to get a proper balance.

Using Compost

Plant Cycle: The normal cycle for a vegetable plant is as follows;

1. Seed sown either direct in growing area, in module trays, seed trays or in see beds for later transplanting.
2. Transplanting were applicable either bare root or a module into final cropping area. All transplants are watered after planting
3. Weeding watering and staking and protecting from pest and disease as needed
4. Harvesting.

Seed Sowing and Transplant Production

Methods of sowing include manual placing seeds either into modules or direct into drills (braodcast sowing used for some spp). There are push seed sowers which will sow and cover a drill much quicker than by hand. Generally though most seed sowing on a garden scale will be manually placing the seed in the drill or tray.

Some still sow seed into seed trays then prick of seedlings into pots or larger trays but with the advent of module trays this practice is seen less.

Pest and Disease Control

Rotation: Avoid diseases especially white rot onions, nematodes in potatoes and club root brassicas.

Beneficials: There are various beneficial organisms that will control pests some can be purchased and released deliberately. This method is most useful in an enclosed area like a greenhouse.

Knowledge of which predators are in your area will give the opportunity to encourage them with cultural methods. e.g. Beetle banks for ground beetles.

Crop covers: Effective against most larger pests including Rodents, Cabbage White and aphids.

Fleece, Mesh and Polythene are the Three main Pptions

Organic sprays: There are various materials which can be used under an organic system, these include soft soap, Ferramol for slugs and copper for potato blight.

Although the organic grower would have to apply for a derogation for the use of some of these.

Cultivation Methods, Weed Control and Tools

Site clearance: If there is a significant covering of weeds these should be first be mown off and either collected then the soil inverted to bury annual weeds. Perennial weeds and roots should be removed by hand before inverting the top soil.

Alternatively a heavy gauge polythene or woven plastic sheet can be laid over the area to exclude light and kill weed growth. This can take up to 2 years if there is a heavy load of perennial weed.

Surface cultivation: Regular cultivation used prior to planting or sowing. Generally it is advisable to cultivate a shallowly as possible to avoid damaging beneficial soil organisms.

Hand tools: single digging with spades, forking over the surface

Mechanical methods: using rotary cultivators, plough.

Sub surface cultivation: Hand tools double digging, Bastard trench which can be used as initial site clearance or to break up a pan.

Mechanical methods using deep tines or sub soiling prior to rotary cultivating or shallow ploughing.

Weed Control after Initial Establishment

Soil covering using polythene or woven plastic membrane prior to sowing or planting: Using this method the soil covering sheet is removed prior to planting or sowing. The objective is to exclude light so preventing germinated weed seedlings from growing. If the ground is covered for six weeks in early spring this can give four weeks before weed seedlings start to grow.

Weed control Mulches. bio degradable and reusable. Laid by machine hand tools or by hand. The crop plants are planted through the mulch so significantly reducing competition from weeds and incidentally reducing soil moisture loss during a dry summer.

Stale seedbed: A method which involves allowing weeds to germinate to seedling leaf stage then hoeing off all the weed leaving a clean bed. The concept is to encourage germination of weed seed in the top layer of soil then hoe off to give a weed free plot for about 2 to 3 weeks.

Weed Control Tools Mechanical

If weeds are being controlled using mechanical methods, hoeing should be done before weeds are established.

Wheel hoe: Hoes dutch, draw and push pull hoes.

- Planting tools.
- Dibber, hand trowels
- Seed sowing tools.
- Seed sower push
- Wolf sower.

Protective Structures

Artificial Lighting/Growing Room: Can be useful early season transplant production to give reliable light and heat when its not available outside. Blue light spectrum needed to produce seedlings.

Polythene tunnels: Larger the better, not normally heated for frost protection, used for soil grown crops, module production. A proper rotation should be used to prevent rotation needed to prevent deterioration of the soil.

Greenhouse: Can be kept frost free, used to raise modules, also useful for growing tomatoes, as due to dry conditions blight is reduced.

Frames/raised beds: Useful for soil grown crops and can be used to bring on transplants. There is the option of changing the cover: polythene to increase temperature, mesh to keep out pests or shade to reduce temperatures.

Loose crop covers fleece and Mesh: Theses are laid directly over the crops, usually with no supporting structures.

Fleece will raise the temperature of the soil and will prevent ingress by pests. They are however very light and can damage easily and get blown about letting in pests which then can breed unimpeded by predators.

Gardening Year

Rough guide to jobs through the year:

Jan Feb.: Seed sowing Alliums& early Lettuce;

- Sowing Carrots in tunnel or frames
- Harvesting winter and tunnel crops

March April.: Cultivations and incorporating compost;

- Mulches laid
- Planting
- seed sowing
- weeding, important to weed down now before the rapid growth stage in May

May June.: Weed control

- Planting, main season for planting, ideal to get most completed by June
- Harvesting early crops
- Seed sowing

July Aug.: Weed Control, weeds are seeding by now so important to remove before seed matures.

- Harvesting
- Planting Latc crops tunnel and frames
- Sowing wintering crops, salads, spinach
- Sowing Green Manures

Sept Oct.: Harvesting.

- Sow Late green Manures
- Late planting in tunnel
- weed control in tunnel
- Sowing broad beans

Nov Dec.: Harvesting winter crops.

6

Vegetables and Fruits in the Fertile Crescent

Vegetables are plants considered fit for human consumption, although they may also double as fodder crops for domesticated animals. As cultural metaphors, they are firmly embedded in all languages and emerge in such expressions as "hot tomato" for an attractive woman, "cabbage head" for someone who is not too bright, or "cool as a cucumber" in reference to extremely calm nerves. However, from a scientific standpoint, the use of the term "vegetable" is highly subjective and is a term of convenience rather than one based on a neatly ordered scientific classification. This ambiguity evolved out of horticultural practice and to some extent out of cultural bias hinging on the key question: what is fit to eat? The vegetable of one culture may be repulsive to another, as, for example, the cannibal's tomato *(Solanum uporo Dunal)* of Melanesia and Polynesia, which was formerly used in salsas for human flesh. On Fiji, the berry was actually cultivated near sites designated for human sacrifice.

Vegetable classification as defined by Western culture derives in part from prescientific attitudes about the mythic world to which those plants belonged in antiquity. Both the Greeks and Romans—and most other ancient Mediterranean peoples—differentiated between two types of edible plants, *holera* or *olera* (cultivated plants) and *horta* (wild plants gathered as food). This dichotomy was presided over by different sets of deities who represented fundamental attitudes about human society and its relationship to nature. We have inherited this structural framework insofar as vegetables are now exclusively defined as oleraceous or esculent herbs, as cultivated in the *hortus oleritus* (kitchen garden). Thus the formal botanical term for growing vegetables is "oleri-culture."

The idea that certain vegetables were suitable for boiling (or conversely that only boiling plants were "vegetables") was highly developed even in ancient Greece. The implication arising from this was that these plants were also worthy of domestication. The ancient Greek physician Diocles of Carystus even went so far as to note that, among the *horta,* beet greens, mallows, sorrel, nettle, orach, iris corms, truffles, and mushrooms were the most suitable for boiling. His beet greens were the ones found growing near the sea, not those cultivated in gardens, presumably the wild ancestor *(Beta vulgaris* L. spp. *maritima)* of the present-day cultivated sorts.

In most romance languages this association with boiling or poaching is further reinforced by such cognates of *oleritus* as Spanish olla (a cook pot or stewing pot), implying that the kitchen garden is designed to produce plants mostly earmarked for treatment in hot water. This connotation is clear in the word *Gemuse,* the German term for vegetables that derives from the medieval German *Gemusz,* a mush or porridge. The idea of cooking or boiling the plants is pre-eminently expressed in the French term *jardin potager,* literally 'a garden for pot dishes: soups and stews.'

The French have also provided English with the word "vegetable." Its most literal meaning is plain: edible vegetation. The accepted origin of the word is that it derives from Medieval Latin *vegetare* via Old French *vegeter* (to vegetate). The classical Latin root is generally given as *vegetus* (lively or active), although this line of linguistic evolution may be subject to revision as more research on French links to Gaulish comes to light. In any case, the old meaning of "vegetate" did not imply something in decline, a common modern connotation, but rather something that was springing to life, a profusion, a natural bounty expanding from seed almost to the extent that it could be heard rustling as it grew—the complete opposite of the gardens of Adonis discussed later in this essay. This concept of vegetal fecundity is quite in line with Celtic ideas about food and nature and is especially characteristic in the obvious lack of definition between the cultivated and uncultivated worlds. Thus, in its root meaning, the English word is more all-inclusive than the Mediterranean concepts of *holera* and *horta.*

This ambiguity is best exemplified by the vast range of plants that are called vegetables in the English language. Tomatoes are berries, yet they are called vegetables. Garlic is considered a vegetable, yet one does not eat the leaves. And what about rhubarb stems, which are treated like a fruit (the leafy part is poisonous), or the sunflower, which is consumed as a seed or as oil? Where does edible vegetation fit in? Most garden books define vegetables as annuals because they are grown to crop for one season only, yet many of these plants are true biennials (celery, carrots, cabbage, leeks, turnips), and some are even perennial (rhubarb, horseradish, sweet potatoes, asparagus, sorrel, peppers).

Another method for defining vegetables was based on the part of the plant considered most esculent for consumption. Here we have the rather simplistic division of the plant world into root vegetables (such as parsnips), pod vegetables (cowpeas, for example), and leaf vegetables (spinach and chard to name two). Again, while it is easy to imagine carrots as root vegetables, lima beans as pod vegetables, and collards as leaf vegetables, the unending discovery of edible plants from exotic locales such as Africa and South America has challenged all Old World definitions of this classification.

In this category is the East African oyster nut *(Telfairia pedata),* a rampantly vining cucurbit that is normally planted at the base of trees large enough to support the heavy fruit. In fact, several vines—both male and female—are grown on the same tree and will produce large squash-like fruit for up to twenty years. But it is the seed, not the fruit itself, that is eaten. Once established, the vines are not given much attention. While oyster nuts are intentionally planted around trees conveniently located near dwellings, this is a vegetable crop based more on passive intervention than on formal horticultural practice. This is a pattern once common in hunter-gatherer societies, where certain patches or stands of useful wild plants were periodically tended and encouraged through selective weeding to produce higher yields.

The yacon *(Smallanthus sonchifolius)* of South America is similarly outside the norm, for it produces crunchy, sweet-tasting tubers that are eaten like fruit. It has been cultivated in the High Andes for so many centuries and reproduced by root division for such a long time that many varieties can no longer produce fertile seeds. They are utterly dependent on humans for survival. Since it is low in calories, the tuber has been recently re-evaluated as a possible diet food.

This continuing botanical revelation has also altered dietary schematics. Old peasant foraged foods like dock have been rediscovered as biodynamic or macrobiotic, synonyms for well-being, longevity, and psychological balance. In this way the vegetable has repeatedly inserted itself into the most radical of lifestyle movements—but this is hardly new, since vegetarianism and the dietary signification of plants has played a key role in philosophical approaches to nutrition since the days of Pythagoras or even since Adam and Eve, the original vegetarians.

Vegetables in the Fertile Crescent

Eden was said to be somewhere in ancient Mesopotamia. About the original garden we know little, but the Mesopotamian peoples left vast heaps of cuneiform tablets that reveal detailed information about the vegetable plants they once cultivated. As the climate grew hotter and drier many thousands of years ago, a vivid visual distinction evolved between cultivated and uncultivated ground. In fact, the vegetable gardens of that era were enclosed by walls both as protection from wandering livestock and as a means to contain and define the area where precious water would be distributed. Furthermore, it is clear from most of the records, whether Assyrian,

Babylonian, or of any of the other cultures sharing the Fertile Crescent, that vegetables were commonly grown around date palms or fruit trees. The palms filtered the blasting sunlight and gave the garden the appearance of a welcoming grove. Thus the garden was also a place where socialization took place and indeed became a place sanctified by its own gods and protective forces.

Archaeological remains (mostly seeds) from this region also support what is documented in clay tablets. The range of vegetables included many still familiar: cucumbers, chards, gourds, onions, garlics, leeks, melons, chickpeas, lentils, cress, kales, and sesame—both for the seeds and for the oil. Colocynth melons, which resemble small watermelons, were grown primarily for medical applications. Some modern Arabic words, such as *kurrat* for a type of leek, can be traced to ancient texts from this period, proof of the long continuity of many of these vegetables. Indeed, even the Arab word for vegetables, *baql,* traces to ancient Aramaic *buqul.*

Tablet inscriptions also point to another feature of vegetable culture that suggests highly developed horticultural practices: specialization. There are numerous references to the "cucumber place" or to the "garlic place," which implies that entire beds were devoted to one type of vegetable and that in many cases this was the sole crop raised by the grower. Garlic was especially valued in this respect and even was used as collateral in financial agreements. In texts where a year-to-year continuity can be reconstructed, it is evident that a garlic place may change into a chard place, so some system of crop rotation must have been in effect. This concept of agricultural specialization was thought to have been perfected by the Phoenicians and codified by Mago, whose great work on agriculture (lost in the original Punic) was highly respected by the Romans.

The Phoenicians were also great middlemen in trading vegetables and seeds throughout the Mediterranean. They are thought to have spread the culture of shallots and artichokes well beyond the eastern Mediterranean, and are known to have introduced the intensive cultivation of saffron into North Africa and southern Spain—saffron was primarily a dyestuff and secondarily a ritual herb. Many Punic words for specific vegetable varieties reflected this plant exchange, as, for example, in *koussimezar* (the Mezar melon or Egyptian "cucumber"), an egg-shaped melon, one of the earliest types to come out of Africa.

Vegetables in Egypt

There is a large body of published material on the history of gardening in ancient and pre-Islamic Egypt, but there are no books per se devoted exclusively to vegetables. Orchards, trees, flora, landscape gardening, and even aquatic plants have received thorough coverage, yet the vegetable stands alone in this curious neglect. Vegetables in general have been viewed as poverty food by most cultures, especially when they form a large portion of peasant diet. Egypt was no different in this regard. The fine gardens of ancient Egypt were enclosed like those of Mesopotamia and contained trees, flowers, even ponds for fish and ducks. The gardens of the peasants were mostly simple agricultural plots devoted to a specific mix of vegetable crops associated with the local economy, as, for example, lentil or onion growing for absentee landlords in large urban centers like Alexandria. On the other hand, vegetable gardens within temple precincts were often quite elaborate and were intended to supply the priesthood with a full range of food as well as offerings for the deity. Lettuce, for example, was important to the cult of Amun-Min, thus its cultivation held both culinary and religious significance. Indeed, temple gardens were considered to be part of heaven, like the temple itself, so vegetables from those gardens achieved a purity unlike those from the common world.

Papyri and tomb paintings have provided a rich array of material dealing with the common vegetables of the day, although not much is known about their actual preparation as food. The most commonly mentioned vegetables were lentils, leeks, lotus, melons, gourds, garlic, asphodel (grown for its bulb), fava beans, chickpeas, fenugreek (ground as flour), garland chrysanthemum (now popular in Asian cooking), cucumbers, onions, lettuce, and mallow. Egypt also served as a conduit for the introduction of watermelons from tropical Africa and, during the late Ptolemaic Period, for the introduction of rice, taro, and sugar cane from Trapobane (ancient Sri Lanka). Due to their dependence on specialized irrigation and cultivation techniques, none of the last three plants spread beyond the eastern Mediterranean until after Arab conquest.

The Greek occupation of Egypt under the Ptolemys radically altered the Egyptian vegetable garden, both with new introductions and in lasting terminologies. *Molokhe* or mallow *(Malva parviflora),* which was once so important to Greek cookery both as a green sauce and as an ingredient in complex recipes, also supplied leaves used like grape leaves for making dolmas. The Egyptians transferred the Greek name to a native wild plant now known as Jews mallow *(Corchorus olitorus),* which was similarly used in green sauces. It is still called *molkhia* in Egyptian Arabic. The use of the same name for plants of a different genus or species is one of the lasting ambiguities inherited from the ancients, who were more apt to lump vegetables together according to how they were used, as in the case of the Roman propensity for treating carrots, parsnips, and parsley root as *pastinaca.*

The Vegetables of Ancient Greece and Rome

There are few surviving writings from the Greeks and Romans that do not mention food and vegetables in some manner. It is known from quotes and citations in works like the *Deipnosophists* of Athenaeus that many books on gardening and agriculture once existed but are now lost. Athenaeus himself lavished considerable attention on foodstuffs, none the least being vegetables. His interests ranged from toasted chickpeas (still a snack food in the Mediterranean) to the medical applications of beets and carrots as vermifuges or a good dish of cabbage to treat a hangover. He even cited the known varieties of lettuce, garlic, fava beans, and many other garden plants in an effort to differentiate which were the best from a connoisseur's point of view.

The surviving Roman work most easily accessible to the general reader is also by another connoisseur, an eccentric called Apicius, whose detailed recipes give specific hints about the role vegetables played in the haute cuisine of imperial Rome. For example, asparagus was baked in eggy casseroles called *patinae,* mallow was commonly added to barley soup, celery made a good stuffing for suckling pig, and turnips marry well with baked duck. There is also scattered advice on when to harvest certain vegetables, as in the case of stinging nettles, which only loose their prickly character when cooked or dried.

Many other works could be cited, such as Columella's *On Agriculture* and especially Pliny's monumental *Natural History,* but the medical writings of the imperial physician Galen are perhaps the richest in detail, since there is considerable commentary on the diet of peasants and farmers and the sorts of vegetables they ate. The aristocratic tone and intended readership of most of the writings that have survived from this period do not provide the kind of first hand observations one might expect from a true master gardener, although Pliny's eye was in fact well nuanced to such details—yet some of his botanical "facts" are obviously scrambled and second-hand. And while it is true Columella certainly knew how to run an estate, he was not the vegetable gardener.

Furthermore, aside from comments about asparagus and cabbage, Cato's treatise *On Agriculture* makes only passing reference to vegetables—be certain to grow them near cities. While this allusion does confirm the existence of well-organized market gardening, Cato's treatise is so loaded with ideology about salty Romanness *(Romanitas)* and the purity of certain rigorous lifestyles that any conclusions drawn from him must be done so with definite reservations. However, it is fairly clear from these and other ancient authors that quite a few cities had developed market gardening to a high degree and even specialized in the cultivation and export of vegetable produce to Rome and other large urban centers.

For example, Cyrene in present-day Libya was well-known in ancient times for its silphium, fragrant saffron, and a mild-tasting tuber now generally identified as taro. It was also the centre for the export of the so-called wild artichoke (*Cynara cornigera* Lindl.), whose domestication was introduced early into Cyprus, Libya, and Carthage from the Levant. This handsome plant is depicted on surviving mosaics in the House of Dionysios at Paphos, Cyprus, and in a mosaic in the Bardo Museum at Tunis. The Roman farmers of Spain and Italy evidently adapted the novel idea of harvesting the flower bud as a delicacy to their local wild cardoons because the artichoke of the western Mediterranean is a subspecies derived genetically from the cardoon (*Cynara cardunculus* L.), not from a wild artichoke ancestor. Buds of the milk thistle, blessed thistle, and safflower were also similarly harvested and eaten.

The dissemination of the artichoke, or at least of the horticultural technology required to cultivate it for food, brings up the larger question of plant exchange during the height of the Roman Empire. Commerce flowed to and from the far-flung provinces in a manner only replicated by the European Union. Archaeology has indeed confirmed that foodstuffs moved quite easily from one place to another, with such exotics as rice turning up in sites in Germany and England. There is also indisputable evidence that, among the aristocracy at least, country life on the great estates attempted to imitate the court life of imperial Rome. Gardens excavated from villa sites have confirmed this.

On the local level, however, Mediterranean cuisine and Mediterranean vegetables were not readily assimilated among the general populace. Joan Alcock's study of food in Roman Britain (2001), based on overwhelming archaeological evidence, reached the conclusion that assimilation was selective, and it was this selectivity that gave rise to the regional cookeries that eventually provided a link of continuity with the regional cookeries of the Middle Ages. This is also the growing consensus of archaeologists in other parts of Europe. Thus, the old saw that "the Romans introduced it" must be requalified, especially since quite a number of vegetables were cultivated in some regions long before the Romans arrived. Cabbage, especially the kales, originated in northern Europe, and in fact, the English words for kale and cabbage are Celtic in origin, as are the German words *Kabbis* and *Kohl,* not to mention the Latin *brassica* and *caulis.* This is evidence in itself that the vegetable exchange during Roman times was complex and two-way, with the Romans themselves learning new things from conquered peoples.

A broader look at Roman literature of all types is especially useful in drawing general conclusions about the state of vegetable gardening from that time. For example, Pliny parodied "gastronomic prodigies," monster vegetables and fruits valued for their size alone. This would suggest that some market gardeners were well acquainted with horticultural practices based on careful seed selection, cold frame techniques, and the manuring of plants at critical periods of growth, much like the modern-day cult of the monster pumpkin or watermelon. There is also a great deal of information concerning value judgments about the role of vegetables in the diet and their cultural significance. Many Roman satires mention garlic as a food only fit for galley slaves and peasants or as something eaten only by soldiers going off to war (garlic heats up the body and therefore creates a warlike spirit). The general drift is that anything flavoured with garlic is therefore rustic and unrefined and as much an antidote to poison as it is to the consuming flames of love (what sweet kiss is not withered by the scorpion sting of garlic breath?). Likewise, fava beans are eaten by jurymen in order to stay awake during trials, their noisy flatulency providing an echoing thunder of divine approval or of legal derision.

The most commonly mentioned vegetables in Roman literary sources include many still known, although in shape and habit they probably did not resemble modern varieties. The list includes turnips, radishes, rocket (arugula to American grocers), leeks, lentils, lettuce, orach, Old World gourds (eaten young like zucchinis), cabbage, onion, peas, chickpeas, fava beans, cucumbers, asparagus, cowpeas, beets, beet chards, sprouting broccoli, watermelon, garlic, mallow, dock, chickling vetch, and blite—otherwise known as purple amaranth *(Amaranthus lividus).*

A number of scholars have taken the liberty of translating *blitum* (blite) as spinach (Gowers, 1996), and this has greatly added to the confusion about the culinary history of spinach because *blitum* is not true spinach. However, something called *barbaricum bliteum* (barbarian blite) also surfaces in Roman literature. This is either true spinach as cultivated by the Armenians and Persians or else good-king-henry, so important to the ritual cookery of the ancient Gauls—the leaves of both plants are similarly shaped. In any case, this mystery vegetable was considered insipid eating and was equated with crudeness and stupidity (insipid people were people who lacked "flavour").

Some of these vegetables also carried a great deal of symbolic baggage, especially in connection with religious cults. Mallow *(moloche)* was considered one of the purest sacrifices for Apollo Genetor at Delos, and Pythagoras himself was said to have lived on it as part of his vegetarian dietary regime because it "washes" the stomach.

Lettuce was associated with the gruesome death of Adonis, the prepubescent lover of Aphrodite, who hid among lettuce before he was gored to death by a wild boar. Thus lettuce was associated with male sterility, effeminacy, and cowardice, and generally was viewed as a suppressant of sexual performance (Detienne, 2000). Gardens of Adonis were planted in pots or baskets during the heat of summer and then allowed to die prematurely on the roofs of houses during the feast of Adonia (19 July during the Roman Empire) as symbolic evidence of the boy's sexual prowess, which produced no "seed" or fruit. Significantly, these little gardens consisted of barley, wheat, lettuce, and fennel, each plant highly symbolic of some aspect of fertility yet a total inversion of what was understood to be garden abundance.

Interestingly, a distinctive lettuce dedicated to Gauas (the Cypriot Adonis), and later known as "Cyprian" during the Byzantine period, was discovered in a Serbian monastery by the U.S. Department of Agriculture in the early twentieth century and is preserved in several American seed banks. It is physically similar to the pointed-leaf lettuce depicted in the medieval *Tacuinum Sanitatis* (Arano, 1976) of the eleventh-century Syrian Christian physician Ibn Botlan. By virtue of this continuity, at least in form and appearance, Cyprian lettuce is a true heirloom variety, a category of vegetable that will be dealt with later in this discussion.

Vegetables in the Early Christian Period

The early Middle Ages is a murky period for the study of vegetables, but a copy (in the Austrian State Library at Vienna) of the *Codex of Dioskorides* dating from 500 to 511 C.E. is illuminated with pictures of plants. The drawings are fairly accurate and convey the important physical characteristics of the vegetables and herbs shown. Thus it is possible to determine that a leek on folio 278 belongs to the Kurrat Group, an ancient type of salad leek mentioned earlier and still grown in the Near East.

The *Codex of Dioskorides* is medical in nature, dealing with the health and dietary aspects of the plants discussed. For a horticultural companion, the *Gheoponika* of Kassianos Bassos, a tenth-century reworking of several older agricultural treatises, provides rules for the cultivation of garden vegetables, thus offering some insights into the seasonal food cycle, both horticultural and culinary, in the old Byzantine East. More specifically, the role of the *Gheoponika* in the provisioning of Constantinople with fresh vegetables has been studied by several historians, most importantly by Johannes Koder (1993). When taken together with the *Book of the Eparch* (prefect of Constantinople) regulating merchants and guilds during the reign of Leo VI (886–912), a relatively detailed picture of market gardening falls into place.

It is perhaps significant to note that by the 1100s many villages in Bulgaria were given imperial privileges that freed them of military duty in exchange for producing food for the court. It is for this reason that the Bulgarians have long been called the gardeners of the Balkans, a status they maintained even under later Ottoman rule.

In the West, the eighth century *Capitulare de villis* of the Holy Roman Emperor Charlemagne is quite valuable for its references to gardens. For example, the *ravocaulos* of that document is believed to refer to a variety of kohlrabi. However, the most priceless garden record is a parchment drawing of the garden plan of the Cloister of St. Gall in Switzerland surviving from the early 800s. It provides a detailed look at how the Roman kitchen garden became transmogrified into a source of both food and medical plants. Sixteen plants are discernible on the plan, including cucumbers, melons, cowpeas, bottle gourds, and smallage (celery resembling parsley). Most important, they are organized into rectangular raised beds. This is one of the earliest references extant to this common garden practice, but it was not unique.

The *Hortulus* of Walahfrid Strabo, abbot (from 838 to 849) of the Cloister of Reichenau on an island in Lake Constance, makes reference to a similar number of plants, again arranged in raised rectangular beds. Strabo's Latin poems about his garden discuss the uses of both herbs and vegetables and is the oldest surviving source on gardening written in Europe during the Middle Ages. Most interesting of all, archaeological exploration of the abbey site has revealed that it was constructed from the recycled ruins of an abandoned Roman villa and that the layout of the garden more or less followed the outlines of the ancient one. The implication is that the Roman gardening tradition maintained by the wealthy during imperial times did not fully disappear at the outset of the Middle Ages. Many estate gardens disappeared completely due to wars and pillaging, but in some regions they simply became fewer in number and passed into non-Roman hands (Percival, 1976). Chateau Ausone near Saint-Emilion in Bordeaux is a famous example of this continuity, although its fame rests on wine not gardens.

The archaeological link is not as clear when it comes to the vegetables themselves, since botanical residues, especially seeds, impose certain limitations on what can be retrieved for science. A carrot seed is indistinguishable from a wild carrot seed and will not tell how the root was shaped or even its colour. Unfortunately, seeds are mostly what one has to work with from medieval sites, although some inferences can be revealing. Cucumber seeds show up in Polish sites in the 900s, thus establishing a bottom line for a vegetable much associated with Polish national cookery. Carbonized fava bean plants from North Germany from the same period reveal that, after the beans were harvested (as winter fare), the plants were used as straw in barns. Seeds, however, do help untangle dates of introduction, and one thing that scholars have learned from medieval archaeology is that the vegetable world was not static, as historians have led us to believe in the past. The broad, flat fava bean, which is the preferred sort for modern cookery, did not appear until the 800s in Spain. This is only one example of vegetable breeding (probably through highly controlled selection) that took place during the Middle Ages, although innovation was indeed slow by present-day standards. Some agricultural historians have suggested that it was the Arabs who created these new types of vegetables—the cauliflower, for example. What can be documented from surviving Arab literary sources is rapid dissemination, but the westward movement of plants in general and vegetables in particular was far more complex than hitherto presumed and remains an area of research ripe for future exploration.

Vegetables and the Arab Diaspora

The agricultural historian Andrew Watson has long promoted the idea that an agricultural revolution took place under early Islamic rule in the eastern Mediterranean, a revolution that was carried westward into North Africa and Spain (Watson, 1983). This has important implications for the movement of vegetable plants. However, other scholarship has questioned this thesis. There is growing evidence that the revolution was already taking place during the late Byzantine period and that it consisted of newer ways of irrigating land and reclaiming marshes so more intensive forms of agriculture could be undertaken. Without entering the question of who invented what, two critical points are undeniable: the technology came out of Persia and South India (the vast irrigation systems in Sri Lanka were well known even to the Greeks), and its spread westward was made possible by the political stability that Arab conquest brought to the regions under its control. It is easy to point to the concentration of wealth in bright spots such as Syria (Damascus and Baghdad in particular), Egypt, and al-Andalus in Spain, but there was an economic implosion in other parts of the newly formed empire. The family papers of the Ibn'Awkal merchants of eleventh-century Egypt reveal a great deal about industrial crops like Egyptian flax or high-profit goods like black pepper, indigo, and sal ammoniac, but information on common garden vegetables is rather limited. That is, unless one looks at medical literature and cookery books. The most heavily used culinary source is also one of the oldest: the *Kitab al-tabikh,* otherwise known as the Baghdad cookery book. It was written down in 1226, although internal evidence clearly indicates that the material was compiled from several much earlier sources, some of which were not Arabic. This ambiguity is one of the difficulties in using cookbooks to pinpoint the introduction of new vegetables. But that said, there are other Arabic cookery manuscripts equally rich in detail surviving from the Middle Ages, such as the *Manuscrito Anonimo* of Moorish Spain, and all of the recipes no matter what the source are fairly clear about the role vegetables played in the diet of the times.

There is certainly no ambiguity in the Baghdad cookery book about the use of eggplants and no doubt at all that the sort discussed had dark black skin (there are directions for removing it). The book also makes ample reference to fava beans, cardoons, rhubarb, leeks, the ridged cucumber (Armenian snake melon), carrots, gourds, taro, cultivated purslane, turnips, sweet fennel, and spinach. There are also references to a form of cabbage commonly translated into English as cauliflower. Without a picture, one cannot be sure (it could be a type of broccoli), but since true cauliflower evidently evolved in the Dead Cities region of northwestern Syria, it is quite likely that this luxury vegetable migrated during the early 800s with its growers when they resettled elsewhere—a small group of those Syrian Christians emigrated to the Karpasia district of Cyprus, where cauliflower was first observed by pilgrims to the Holy Land later in the Middle Ages. The cauliflower is not mentioned in European scientific works until specimens are discussed by Dodonaeus in 1560. By that time Cypriot seed was being exported to northern Europe via Venice.

One is also treated with a rich array of vegetables in another work called *Kitab Wasf al-Atima al-Mutada* (Description of Familiar Foods) written in 1373. Of particular interest is the differentiation of several types of leeks, indicating not only distinct varieties but also distinct culinary uses at different stages of development—indirect evidence of a highly evolved sense of market gardening. Four sorts of leek are mentioned: the vegetable leek *(kurrath baql),* the Nabatean leek, the table leek, and the Syrian leek. The first is not a variety but rather a spring leek, young greens similar in character to Chinese garlic chives. The Nabatean leek may be equated with the modern salad leek of Iraq, a member of the Kurrat Group, short in height and rather deep-rooted. The table leek is a blanched leek similar to the Catalan Calcot onion, and indeed the cultural technique of burying them in deep trenches may be the same. The Syrian leek is the *kephaloton* of medieval Cyprus, a Greek word derived from Syrian *quaflot,* a leek with an unusually large bulb. This plant is the genetic ancestor of the modern elephant garlic. Under the name *Porrum Syriacum,* it was first illustrated in European botanical literature by Tabernaemontanus in 1588.

Arab books dealing with cookery exhibit an undeniable passion for elegant preparations, even with simple vegetables. But such food was the privilege of the aristocratic few, and the wealth that sustained that lifestyle was soon to fade with the economic upheavals caused by the Crusades. Until the late twentieth century, historians have greatly underestimated the exchanges that took place after the establishment of Latin footholds in the Levant and Byzantium or the role of large Christian minorities that persisted in Egypt and Syria during the early Muslim period.

The Nestorian or East Syrian Church, which spread into Persia, established strong trading communities in China and Malabar as well as in Cyprus, where the Nestorian Lakhan family became extremely powerful based on trade in medical aloes. One line of trade and plant exchange went through Tabriz in Persia overland to the Caucasian kingdom of Georgia and the Greek Empire of Trebizond on the north coast of modern Turkey, all to circumvent the Arabs. That this Black Sea entry was an important route for the movement of Asian food plants westward may be inferred from an eleventh-century Byzantine reference to the "citrons" of Anatolia, a variety of lemon introduced via Georgia and Armenia and still preserved in Georgian botanical collections. Eggplants also followed this route.

European contact with foods of the Arab world was not limited to the crusading troops that went to the Holy Land and returned. The Latin Kingdom of Jerusalem (1099–1291), the Principality of Achaea (1205–1430), a French feudal state established in Greece with its capital at Mistra, a Catalan principality centred on Athens, various Venetian and Genovese ports, and the sister kingdoms of Cilicia (1080–1375) in Asia Minor and on Cyprus (1192–1489) were all characterized by colonial aristocracies with highly orientalized foodways.

The last kingdoms, especially that of Cyprus due to a Papal Bull, served as conduits for the spice trade with the Muslim world. In the case of Cyprus, the kingdom lent its name to an international style of cookery mentioned in numerous medieval cookbooks. More important, the intermarriage of wealthy Latins in the Levant with European nobility, particularly with families in Aragon and in northern Italy, brought to Europe a constant influx of personal cooks, gardeners, and retainers schooled in eastern Mediterranean ways. It is not surprising that some of the earliest references to exotic vegetables like eggplants, cauliflowers, okra, and numerous sorts of Near Eastern melons show up in late medieval Italy.

Vegetables in the Renaissance

The *Carrara Herbal* (British Library, Egerton MS 2020), created sometime before 1403, was one of the first late medieval herbals to depict plants and vegetables accurately, although it may have been based on a now lost Byzantine prototype. Such illustrated handbooks of health, as well as numerous herbals, offer a rich visual record of the sorts of vegetables deemed. The beautiful gardens witnessed by travellers through the Latin East were now replicated in Italy but with the goal for reattaining a glorious Roman past. The Italian pleasure gardens of this period so impressed Casimir the Great of Poland that he installed one in Cracow during the 1360s, complete with cold frames for forcing Mediterranean vegetables. In short, the vegetable garden once again becomes an object of status.

The discovery of printing, followed quickly by the discovery of the New World and the heady harvest of its vegetable riches, only accelerated a quest for new and exotic things to ornament the gardens of the rich and powerful. Tomatoes, peppers, potatoes, sunflowers, beans, sweet potatoes, new sorts of pumpkins, and a new kind of wheat called maize fill the pages of botanical treatises and plant books of the period. The 1500s may be characterized as a century during which botanists attempted to organize the vegetable world into some type of scientific order, although that "order" by modern standards was quite chaotic. For example, a confusing observation is that the Jerusalem artichoke from North America (not from Jerusalem and not an artichoke) is known as *Flos solis Farnesianus* (Farnese sunflower) in reference to the fact that the gardens at the Villa Farnese in Rome provided several botanists with the first known specimens. Sorting out such conflicting nomenclature has plagued garden historians ever since.

However, botanical gardens were established in this century, the first in 1545 at the University of Padua, and some of the greatest botanical works of the Renaissance were issued during this era, especially those devoted to cataloging the gardens of such important plantsmen as Conrad Gesner in Switzerland (1561), Georg Fabricius in Meissen, Germany (1569), and Camerarius in Nurnberg (1588). All of these books contain valuable woodcuts depicting vegetables, and many medieval favourites like skirret *(Sium sisarum)* and monk's rhubarb *(Rumex patientia)* are shown for the first time. Vegetables also figured prominently in Renaissance art and paintings, especially the still life genre. Among the most whimsical vegetable compositions are those by the court painter Giuseppe Arcimboldo (c. 1527–1593), who used vegetables and fruits to create faces and other conceits.

The most significant body of literature, however, was the garden guides that discussed not only specific vegetable varieties but also how to grow them. The French work known as *L'agriculture ou la maison rustique,* first published in Latin (1535) by the Paris printer Charles Estienne, was soon translated into most major European languages. Marco Bussato's *Giardino di agricultura* (Venice, 1592) was also extremely influential, as was Johann Coler's *Oeconomica ruralis et domestica* (Wittenberg, 1597).

The great classic, however, was Oliver de Serre's *Theatre d'agriculture* (Rouen, 1600), which became a standard garden book for much of the next century. This great outpouring of garden knowledge was capped in many ways by the lavishly illustrated *Hortus Eystettensis* assembled by Basilius Besler in 1613 for his patron plant collector, the bishop of Eichstatt in Bavaria. Not only did the good bishop own prize specimens of rare eggplants, balsam apple *(Momordica balsamina),* tomatoes, and domesticated asparagus, his potted prickly pear cactus from the New World required a wooden superstructure to hold the monster plant in place.

Vegetables in the Baroque Period

The seventeenth century witnessed a revolution in botanical science and the proliferation of books devoted to illustrating plants and vegetables from many parts of the world, including new introductions from Asia and the Americas. Francisco Hernandez's *Nova Plantarum* (Rome, 1651) was devoted almost exclusively to the foods of Mexico and included native names for the plants. His woodcut illustrations offer a priceless look at the characteristics of common vegetables then grown in New Spain, including the lowly miltomatl or tomatillo *(Physalis ixocarpa)* now popular in Southwest American cooking.

Other books containing strange designations like *Pomum amoris majus fructu luteo* (large yellow fruited love apple or, more simply, yellow tomato) remind one how much lack of order prevailed in the scientific naming of newly discovered vegetables and how much has changed since Linneaus imposed order on the world of plants in the eighteenth century. Names like *bamia Aegypitiaca* (Egyptian okra—*bamya* is actually a Syrian word) of one author might become "ladies' fingers" of the next. Likewise the *lactuca hispanica* (Spanish lettuce) of one author was the *Cos* or Roman lettuce of another, the names more often than not reflecting the source of seed rather than the true origin and history of the vegetable. One of the most fashionable cabbages of the period was the so-called *Brassica tophosa,* better known as black Tuscan palm tree kale, "rediscovered" by American seeds-people under the new moniker "dinosaur kale." The penchant for fanciful names has not changed.

If a generalization can be made about the seventeenth century, it is that the rare and exotic vegetable of the previous century gradually became the daily fare of the urban middle class by 1700.

Plant breeding, especially in Holland, brought many new sorts of vegetables onto the market. Named varieties of potatoes, carrots, celery, chicory, peas, and turnips soon proliferated in kinds and colours. Added to this roster were newly discovered

Asian foods, like Malabar spinach *(Basella alba),* introduced from Java in 1688. Handbooks on plant breeding were even published, one of the earliest in English being Walter Sharrock's *History of the Propagation and Improvement of Vegetables* (1660). From this time on, the vegetable undergoes a steady refinement with emphasis on greater delicacy of flavour, more beautiful shape, and increasing tenderness.

Much of this was directly connected with shifts taking place in cookery, especially the use of vegetables in sauces and elaborately prepared dishes. Vegetables were also given ornamental value with paring knives, so turnips feathered out into birds, carrots unwound into golden fish, and the cookbooks of the day are full of illustrations showing how to do this. Most notably, however, the vegetable became a prized market commodity; growing of vegetables, a respectable line of work for the honest laborer; and period depictions of market scenes never fail to convey the impression that only the best has been laid before the eye.

Aside from shifts in cookery, the virtues of country life and the pursuit of its simple pleasures helped elevate vegetable gardening as a worthy and genteel pastime. Jan van der Groen's *Den Nederlandtsen Hovenier* (Amsterdam, 1669) was extremely influential in this respect, as were Nicolas de Bonnefon's *Les delices de la campagne* and *Le jardinier francois* in France.

All of these works were translated into several languages and included specific discussions of vegetable varieties and cooking tips. Under fava beans, for example, Bonnefon recommended several different methods of preparation, including fricassees like peas or boiled plain with slices of bacon, noting that fresh green savoury went "marvelously well" with any fava bean dish.

Not the least important, from the standpoint of vegetables, was John Evelyn's *Acetaria: A Discourse on Sallets* (London, 1699), which discussed most of the popular types of vegetables of the day, especially ways to employ them raw or semicooked in salads. Of earth chestnuts *(Bunium bulbocastanum)* he remarked: "the Rind pard off, they are eaten crude by Rustics, with a little Pepper; but are best when boil'd like other Roots, or in Pottage rather, and are sweet and nourishing."

Vegetables in the Enlightenment

The idea that vegetables recaptured the original wholesomeness of Eden became an underlying theme for many of the more offbeat cookbooks of the eighteenth century, with *Adam's Luxury and Eve's Cookery* (London, 1744) considered one of the most typical. The underlying philosophies expressed in these books may be said to represent the intellectual forerunners of true vegetarianism, which was indeed practiced in colonial North America by the so-called White Friends, a group of Quakers who wore clothing of unbleached cloth. The most highly organized vegetarians in early North America, however, were the Bible Christians, who expanded from England in 1816. Martha Brotherton's *Vegetable Cookery* (London, 1833) became the dietary handbook for this group. The European penchant for country life was quickly transferred to England during the 1600s and from there to colonial America. Doubtless it achieved its American apotheosis in such famous estates as William Penn's "Pennsbury Manor" along the Delaware River, Thomas Jefferson's "Monticello" in Virginia, William Hamilton's "Woodlands" near Philadelphia, and Charles Carroll's "Mount Clare" in Baltimore. Jefferson's personal garden account book, published by the American Philosophical Society in 1944, remains a lasting testimony to the central role that kitchen gardens—and vegetables in particular—played in this manorial lifestyle. It was on such estates as these that many of the Old World exotics were first introduced to North America. Charles Norris of Philadelphia, for example, is known to have raised black-skinned eggplants, since a letter survives from 1763 imploring him for seed. Many of the most popular vegetables of this period can be found in Philip Miller's *Gardener's and Botanist's Dictionary* (London, 1759), and remarkable as it may seem, some of Miller's vegetables are extant, among them red celery, Spotted Aleppo and Silesia lettuces, spinach beets, and domesticated sea kale.

The Vegetable in the Nineteenth Century

The Industrial Revolution in England and Europe created the need for a new type of gardener and, indeed, new sorts of vegetable varieties. Market gardening had always existed to some extent in and around urban centers, but the huge new concentration of landless workers packed into the cities meant someone else would have to act as a surrogate kitchen gardener to supply their tables. Thus vegetable horticulture underwent rapid specialization with growers focusing on such basic food crops as cabbages and onions, or turnips and potatoes.

New demands were placed on vegetable breeders for vegetables that would travel well and that gave a good appearance even after rough handling. The old-time requirements of the country kitchen gardener for vegetables that dried or pickled well were overthrown by cold frame and hot-house horticulture that could deliver such tender things as fresh peas and lettuce all winter. Furthermore, most of the old medieval vegetables like skirrets, rampion, orach, sow thistle, and nettles dropped out of mainstream diet and by the mid-nineteenth century were largely associated with rural poverty, those "rustics" mentioned by Evelyn in his treatise on salads. So thorough has industrialization distanced the consumer from the *horta* of old traditional northern European diet that the wild greens of Mediterranean culture, indeed any unusual food plants from Asia or Africa, are embraced as an antidote to a diet adrift of its "natural soul."

The two bibles of nineteenth-century American kitchen gardening are arguably Bernard M'Mahon's *American Gardener's Calendar* (Philadelphia, 1806) and Fearing Burr's *Field and Garden Vegetables of America* (Boston, 1863). Both of these works were reprinted in the twentieth century because they contain long, valuable lists and detailed descriptions of many nineteenth-century vegetables. Burr even included small woodcut illustrations. M'Mahon's book is a cultural bridge with the past, for it contains much on the old types of vegetables still eaten by traditionalists of his day. His is also a book written for the person who gardens, in particular a person who employs a staff, thus his vegetables fall into that category of food destined for genteel tables. Fearing Burr's book was quite another sort.

Burr's primary interest was to catalog the best commercially available vegetable varieties so gentlemen farmers and market gardeners would be guided in their selection of the best seed for the best investment. The book was arranged like a scholarly encyclopedia, and it was not cheap, two points against it from a farmer's standpoint. Its New England stodginess was quickly overtaken by Peter Henderson's *Gardening for Profit* (New York, 1865), an inexpensive handbook that became a best seller and the blueprint for true truck farming. Henderson was a seedsman, and he was not blind to the fact that small-scale farming for urban markets would require yet another type of vegetable; indeed, his book marks the birth of the so-called commercial vegetable grower.

Henderson's effect on vegetable breeding was immediate, and nowhere in the United States did it manifest itself more than in the explosion of newfangled high-profit tomatoes. The Paragon tomato was introduced in 1870 along with the canning jar of the same name—the connection was not coincidental. The leading tomato breeder of the time was Alexander Livingston (1822–1898) of Ohio. His Acme tomato (1875), Golden Queen (1882), Beauty (1885), and Stone (1891) are still grown and are considered among the classics of American commercial garden vegetables. But Livingston was only one of a number of seeds-people actively engaged in creating newer and better vegetables for the market. Perhaps the dean of them all was W. Atlee Burpee of Philadelphia, who even went so far as to offer large monetary rewards for backyard discoveries worthy of commercialization. Many of Burpee's best vegetable introductions came from housewives and farmers who happened to have an eye for the unusual, of which the Montreal Market Melon is a prime example. Burpee's sense of marketing was also shrewdly sophisticated, because he hired Sarah Tyson Rohrer, owner of the Philadelphia Cooking School, to create a cookbook called *How to Cook Vegetables* (Philadelphia, 1891), thus insuring that even the most helpless beginner would feel confident in buying his seeds. After all, they received the cookbook as a bonus.

The second half of the nineteenth century witnessed a large increase in the number of vegetables developed solely for manufacturing purposes not only in the United States but in Europe as well. In France, the tiny cornichon pickles were introduced in the 1880s. In England, the Marrowfat pea became the ubiquitous pea of the canneries, both as a canned vegetable and as pea soup. Whole communities grew up around the production of one vegetable. In Germany, Zittau became famous for its monster onions and Tetlow for its dwarf turnips; Poitou in France became famous for its golden leeks. In the Austro-Hungarian Empire, at Znaim in what is now the Czech Republic, a pickling industry was established in 1852 based entirely on the Znaim cucumber, a mutation of a cucumber brought from Asia in 1802. This pickle capital of central Europe remained in business until 1945, when it collapsed under ethnic cleansing. Today the Znaim cucumber is extinct.

The need for fresh breeding stock to supply growers with newer and better varieties of plants lay behind the establishment of several experimental gardens in the late 1800s. The most famous was at Crosnes (Seine-et-Oise) in France, a place-name now attached to a tuberous vegetable *(Stachys affinis)* from China first trialed there in the 1880s. The heads of that garden were Auguste Paillieux (1812–1898) and his assistant Desire Bois. The two of them coauthored a garden classic known as *Le Potager d'un curieux* (Paris, 1885), and Bois later published his own master-work, *Les Plantes alimentaires* (Paris, 1927). These men were responsible for a large number of new vegetable introductions, and they were interested as well in their histories. They valued the research of such botanists as Emilii Bretschneider, whose history of plant exploration in China (1898) remains one of the milestones of nineteenth-century horticultural writing. They were also keenly aware of the value of bio-diversity.

The greatest collector of vegetables and edible plants in general specifically for their genetic interest was Nikolai Vavilov (1887–1943) of Russia. The Vavilov Institute, founded at St. Petersburg in 1905, became a model for similar gene banks established in other countries, including the United States. Vavilov's collections gathered from all parts of the world are considered priceless, and they have taken on much greater importance since the development of hybrids and genetically engineered foods.

In the past the hybrid vegetable was viewed as a worthless mule, and most growers disdained them because they were not fixed in their characteristics, thus they would not grow true from seed. A revolution took place in this thinking during the 1940s as seed companies began to promote the benefits of controlled crosses to yield vegetables with specific traits. Such plants are known as F1 hybrids, the label standing for "first generation." This concept has further evolved with the idea that the precise genetic mix to create that plant may be patented and thus owned in perpetuity. Genetic engineering has added one more element to the mix: traits borrowed from near species or from other life-forms to create vegetables that would not have occurred in nature. This type of vegetable dominates agribusiness, but it has also come under attack from many quarters under the moniker of "Frankenfood" (a word combination of "Frankenstein" and "food").

The primary argument against the hybrids is that, since seed cannot be saved, growers are forced to purchase new seed each year. For large-scale farmers, this increases the debt side of the ledger at the cost of seed-saving convenience. Another argument is that hybrids on the whole lack the same vigor as open-pollinated plants, that there is some long-term genetic decline at play. The patented vegetable has been criticized on moral grounds (who owns Nature?) and because it is easy to circumvent the patent by creating renegade varieties with slightly different genetic compositions. Furthermore, the ownership issue resembles the sort of complex economics of old tithes and quitrents that eventually changed medieval agriculture from a tenant system to serfdom. This New Feudalism is based not on class privilege but on farms mortgaged to banks, fertilizer companies, and seed suppliers.

While the pros and cons of these arguments are primarily limited to the farming side of the equation, it is an issue that ultimately affects the consumer and the price of vegetables on the shelf. These issues are also more polarized in developed countries like the United States than in Africa, Asia, or Latin America, where small-scale farming is the norm, and the economics of farming are different and are not necessarily based on cash flow or cash crops. However, the financial success of hybrids, patented varieties, and genetically engineered food is predicated on cheap oil for transportation, not to mention the chemical by-products of oil translated into herbicides and pesticides required to maintain these specialized vegetable crops. Oil, not politics or environmentalism, may ultimately determine the future of such vegetables. But there is also another revolution taking place. It is known as artisanal agriculture, its technology is increasingly organic, and it is based on open-pollinated vegetables, many of which are also heirloom varieties.

The Heirloom Vegetable and the Ten-Acre Farm

The discovery of vitamins and the role they play in human nutrition raised the status of the vegetable from an adjunct to the meal to a much more centralized role. When it became clear that legumes could deliver much of the same nutritional value as meat, a new look at all sorts of vegetables seemed to be in order. On hindsight, the appearance of Eleanour Sinclair Rohde's *Uncommon Vegetables* (London, 1946) and John Organ's *Rare Vegetables* (London, 1960) should have served as warnings that a shift was taking place in the food world and that this shift was the harbinger of something new. Several elements appear to have converged and to have begun working in conjunction with one another.

The first was a general setting of rapid economic growth, especially in the 1980s, and the flow of some of this new-made wealth into dining out. The second was the mainstreaming of vegetarianism and more broadly of healthy lifestyles, which placed new demands on the market (and on restaurants) to come up with a more challenging and nutritionally satisfying range of vegetable choices, although not necessarily inexpensive choices. The third was the American Bicentennial in 1976, which spawned a renewed look at historical foods and foodways and which gave rise to a number of grassroots organizations devoted to studying and preserving foods of the past. The most important plant organization, and one with enormous continuing influence, is

Seed Savers Exchange in Decorah, Iowa, the high temple of what is known as the heirloom vegetable.

The fourth and perhaps the most significant development was the growth of small or artisanal farms devoted to supplying high-end urban markets with the vegetables demanded for a new American cuisine variously styled California cuisine and championed by such activist restaurants as Chez Panisse in Berkeley, California, and the White Dog Cafe in Philadelphia. The core concept of California cuisine was the use of locally grown food products with an emphasis on freshness, originality, and organic horticultural methods. This type of model could be replicated anywhere in the country, and thus a real interest in heirloom vegetables was born.

The heirloom vegetable is a variety that has been handed down from the past. It may be an old commercial variety like Conover's Colossal Asparagus (1863), a garden classic like Victoria Rhubarb (1837), or a symbol of American gardening genius like the Brandywine tomato (1889). Whatever their role, these are plants that have been preserved more or less intact since they were developed. Their historical and cultural genealogies may be impressive, and this alone appeals to many people who sense a loss of cultural identity and who are looking for a means to recover their "natural souls." Native American heirlooms are extremely potent in this regard.

From an economic standpoint, the heirloom vegetable is free of the ownership issues inherent in hybrid, patented, and genetically engineered food. The heirloom belongs to the community, and an exchange of this food (as well as its seeds) is viewed as a strong link in the commensality of people and a link with nature, especially since the plants are pollinated by natural means out under the sky. On a less philosophical level, heirloom vegetables were developed to meet the agricultural needs of specific soils and environments and therefore do not require the same economic investment in fertilizers and insecticides as hybrid varieties. They are also strengthened by an inherent genetic diversity lacking in hybrid sorts—a built-in mechanism to prevent massive crop failure. This appeals to organic growers, and the surprisingly rich flavors of plants raised in this manner have gained many converts in the food community at large. Politically, the heirloom vegetable represents an alternative to the type of market control and genetic expropriation associated with agribusiness. In developing countries, this means local control of local food resources, a theme championed by the Food and Agriculture Organization of the United Nations.

When the astronauts first set foot on the moon, they looked back to behold a blue orb in the darkness of space. That view changed humanity, because it said as no words could express that Earth is indeed an Eden and perhaps, for all its vicissitudes, our only reward. The philosophies that have guided us in the past are seriously challenged. While it is true that vegetables are not perceived by humans to react to pain, fear, or anxiety and while they do not have red blood, who is to say they are not our protectors?

7

Vegetable Research and Management

Development of a large number of improved varieties and wider adaptability and standardisation of their production technologies for various agro-climatic conditions has made it possible to produce vegetables in wider areas and has improved the prospects of their supply tremendously as follows.

(i) *Garden pea:* variety 'Arkel' has revolutionised the production of early peas in all pea growing areas.

(ii) *Cauliflower:* variety 'Pusa Early Synthetic' has adapted to warm climatic conditions of Tamil Nadu and has made it possible to grow cauliflower commercially in this non-traditional area.

(iii) *Watermelon:* variety 'Sugar Baby' has spread fast in entire Northern and Eastern India and has benefitted both the growers with better remuneration and the consumers with better remuneration and the consumers with superior quality. Another variety 'Arka Manik' has made a dent in the Southern and South-Western parts of the country.

(iv) *Okra:* variety 'Pusa Sawani' bred for resistance to yellow vein mosaic virus prone areas/seasons replaced all other local varieties from cultivation all over the country.

(v) *Tomato:* variety Pusa 'Sel-120' has made it possible to achieve high yields of quality produce in root-know nematode infested soils. With the released of cold tolerant variety 'Pusa Sheetal', we can now grow tomatoes all the year round.

(vi) *Radish:* with appropriate choice of suitable varieties for specific seasons now we can grow radish round the years.

(vii) *Onion:* Until 1978, Kharif onion cultivation was only grown in Maharashtra, Gujarat, Andhra Pradesh and Tamil Nadu. However, identification of variety

N-53 and ADR and development of technology for kharif onion has enabled to get two crops of onion annually in Northern India where it used to be only a winter/spring crop.

(viii) *F1 Hybrids:* There has been an alroere has been an alround appreciation of growing of F1 hybrids in vegetable crops. Sizeable area of 20,000 ha. is estimated to be covered under tomato hybrids in Karnataka, Maharashtra & Southern Gujarat. Similarly the area under F1 hybrid of cabbage is estimated at about 8,000 ha.

Gaps and Constraints

Despite a large number of varieties and hybrids developed, the productivity of vegetable crops has not improved. Varieties with longer shelf life and suitable for processing are very few. Multiple disease resistant varieties are yet to be developed. Although a lot of work on heterosis breeding in vegetable crops has been done in the country, yet there is acute dearth of real good hybrids in crops like tomato, cabbage, cauliflower, onion etc.

Vegetable based cropping systems have not been developed so far, which puts these crops at a disadvantage for fitting them in crop rotations. Excessive use of pesticides has created problems of pesticide residues and hence there is a need for integrated pest and disease control.

Briefly, these gaps are identified as follows:

1. Insufficient germplasm in different vegetable crops and the need to augment indigenous and exotic-collections.
2. Lack of new high yielding vegetable varieties/hybrids, carrying high degree of resistance to diseases, pests, environmental stresses etc.
3. Absence of basicts, environmental stresses etc.
4. Absence of basic/fundamental research programmes to accelerate the work on applied aspects.
5. Lack of competent scientific manpower to manage the research programmes.
6. Lack of infrastructure and fund availability for research on vegetables.

Future Research Priorities

While significant progress has already been made there are still several kproblems to be tackled. For this, the following research priorities have been identified.

1) Breeding for resistance to abiotic factors *viz.* diseases and insect-pests such as:
2) Breeding for resistance to abiotic stresses eg. salinity, alkalinity, salt tolerance and stress environment *e.g.* hot set and cold set tomatoes.

3) Heterosis breeding in onion, tomato, cabbage, cauliflower, cucurbits, brinjal, and capsicum.
4) Breeding for nutritional and processing qualities in vegetables like tomato, onion, peas and garlic (dehydration).
5) Use of biotechnology for incorporation of resistance to disease/pests/abiotic stresses.
6) Intensification of research on seed production of temperate, tropical and sub-tropical vegetables and intensification of breeders seed production programme.
7) Export oriented research on vegetables like onion, chillies, okra, peas, tomato, brinjal, cucumber, cauliflower and cabbage.
8) Developing efficient cropping systems.
9) Research on growing vegetables in protected environments.
10) Research on off-season vegetable production and under exploited vegetables.
11) Studies on insecticidal residues.

Tomato	leaf curl virus, TMV bacterial wilt,phytopthora blight, fruit borer.
Brinjal	fruit and shoot borer, bacterial wilt, little leaf.
Okra	yellow vein mosaic and pod borer
Chillies	virus and pest complex
Onion	purple blotch, stemphylium (moth and thrips).
Cucurbits	downy mildew, powdery, CMV, fruit fly.
Cole crops	Sclerotinia, Alternaria and soft rot.
Peas	Powdery mildew
Beans	Septoria, mosaic virus and b
Beans	Septoria, mosaic virus and bruchus

Plant Nematology in Developing Countries

This short contribution introduces the major nematode pests in developing agricultural systems and discusses possible measures for their control.

*Root-knot Nematodes (*Meloidogyne *spp.)*

Meloidogyne species constitute the major nematode problem in developing countries. The most common species is *M. incognita,* which causes considerable losses in many crops. Many vegetable crops, in several botanical families, such as Solanaceae, Cucurbitaceae, Leguminosae, Liliaceae, Chenopodiaceae, Compositae, Umbelliferae, Cruciferae and Malvaceae, suffer the greatest damage. But *M. incognita* is also highly pathogenic to some staple crops such as cereals, including rice, maize, potato, soyabean, banana, plantain, sweet potato and yam; or industrial crops such as tobacco, coffee, sugar cane, sugar beet, cotton and black pepper. Often this species also causes economic damage to fruit crops such as guava, pineapple, papaya and grapes.

Meloidogyne javanica is also widespread and prevalent in warmer climates. It attacks almost all the plants affected by *M. incognita* except cotton and pepper, although in the case of pepper there are a few reports. Many vegetable crops are severely affected by *M. arenaria* which, however, causes the most significant losses on groundnuts, especially in West Africa. *Meloidogyne hapla* has a more restricted distribution, being more common in cool climates or on winter crops in the Mediterranean area. Major economic damage has been reported on carrots, potatoes, alfalfa, clover and strawberries.

A few other species of root-knot nematodes have restricted areas of distribution and affect locally various crops to different extents. Among them is *M. artiellia,* pathogenic to legumes and cereals in southern Europe and in the Near East. *Meloidogyne naasi,* which is widespread in northern Europe, has been reported affecting cereals in South America. *Meloidogyne exigua* and *M. decalineata* attack coffee in South America and Africa, respectively. *Meloidogyne brevicauda* is pathogenic to tea planted at high elevations in Sri Lanka. *Meloidogyne graminicola* affects cereals in India.

Cyst Nematodes (Globodera spp., Heterodera spp. and Punctodera spp.)

Less cosmopolitan, but as pathogenic as root-knot nematodes, are species of cyst-forming nematodes. Among them the most widespread are certainly *Globodera rostochiensis* and *G. pallida.* They are very common in Central and South America where they are thought to be indigenous, but their occurrence has now been reported also in hilly African and Asian countries. They cause considerable damage in the western countries of North Africa.

Heterodera avenae is a common species in temperate areas of India where it is the causal agent of a disease of wheat called molya. It also attacks barley. *Heterodera schachtii* attacks sugar beet in North Africa, the Near East and Mexico. *Heterodera sacchari* is reported to damage sugar cane in Africa and India and rice in Africa. *Heterodera oryzae* parasitizes rice and banana in Africa. *Heterodera cajani* is perhaps the most pathogenic species in India on various leguminous crops such as pigeon pea, cowpea and chickpea. *Heterodera zeae* is very widespread and affects maize, sugar cane and sorghum from North Africa to India. Of local importance, although destructive, are *H. daverti* and/or *H. trifolii* in North Africa. *H. ciceri* and *H. latipons* are other important nematode pests in the Near East, *and Punctodera chalcoensis* in Mexico.

Other Important Nematode Pests

Anguina tritici is a common parasite of wheat on the plateaus of Africa and Asia. In Asia and, to a lesser extent, in Africa, *Aphelenchoides besseyi,* the causal agent of the white tip disease of rice, is responsible for major losses of this cereal. *Aphelenchoides ritzemabosi* is widespread in hilly areas of northern India on zinnia and *A. agarici* seems to be a destructive pest of cultivated mushrooms in India.

The stem nematode, *Ditylenchus dipsaci,* attacks several crops in the Mediterranean region and causes major losses of onion, garlic, broad bean and strawberry crops. *D. dipsaci* is also a major pest of garlic in many countries in South America. The phytopathological importance of *D. angustus* to rice in the Far East is also well known.

Within the genus *Helicotylenchus, H. dihystera* has worldwide distribution, being a pest of many crops. Its damaging effect, however, has been demonstrated only on olive in Egypt and on rice in Mauritius. *Helicotylenchus indicus* causes some damage to vegetable crops in India, but only when at high population densities and *H. multicinctus* occurs wherever bananas are cultivated. Fortunately, this species is a weak parasite affecting only the cortical tissues and causing economic damage only at high population densities. *Mesocriconema onoense, M. curvatum* and *Hemicycliophora typica* are parasites of rice in Africa.

Various species of *Pratylenchus* parasitize crops in developing countries. *Pratylenchus brachyurus* attacks many plants in Africa and in the Mediterranean region. Under tropical climatic conditions *P. coffeae* is an important pest of coffee, banana, cowpea and citrus. *Pratylenchus penetrans* is frequently associated with declining date-palms in Saharan oases. *Pratylenchus thornei* and *P. mediterraneus* are parasites of legumes and cereals in North Africa and the Near East and. *P. loosi* is a major pest of tea in Sri Lanka and India. Finally, *P. zeae* attacks maize and sugar cane in Africa and Asia.

Radopholus similis is a destructive parasite of banana throughout the tropics. It also causes great damage to black pepper in Java, to coconut palms in India and to tea in Sri Lanka.

Species of *Hirschmanniella* such as *H. oryzae, H. gracilis, H. spinicauda* and *H. mexicana* cause varying extents of damage to rice in Africa, Asia and Mexico.

Hoplolaimus seinhorsti has been shown to be pathogenic to black pepper and various vegetable crops in Sri Lanka, and *H. indicus* and *H. pararobustus* are often found associated with a decline of sugar cane and other plants in Africa and Asia. Yams in Africa are severely affected by *Scutellonema bradys,* and it is also an important Post-harvest pest attacking yams in storehouses.

Hemicriconemoides mangiferae may be the most widely associated nematode species with mango. It occurs mainly in Asia and Africa where it is found also in the rhizosphere of other fruit-trees. Among *Tylenchorhynchus* species, *T. brassicae* is the most frequently occurring in the rhizosphere of crops. It is associated with different plants in Africa and Asia. Other common species are *T. claytoni* and *T. indicus.*

Rotylenchulus reniformis is widespread in all the regions with tropical and subtropical agriculture. It causes economic damage to many plants, including vegetable crops, fruit-trees and staple and fibre crops. The citrus nematode, *Tylenchulus semipenetrans* is invariably associated with citrus groves, which are affected to different extents depending on the paedoclimatic and edaphic situations. *Rhadinaphelenchus cocophilus,* the causal agent of red ring disease, is very destructive on coconut and oil-palms in Central and South America.

There are species of *Longidorus, Paralongidorus* and *Xiphinema* that may cause serious problems, although restricted to local situations. *Longidorus africanus* is a parasite of cotton in Mexico and the Sudan, but it also occurs in the Near East and East Africa. *Longidorus laevicapitatus* attacks sugar cane in many African countries, *Longidorus cohni* is a pest to cereals and alfalfa in the Near East, *Paralongidorus buchae* attacks many crops in Mauritius, *P. oryzae* is a parasite of rice in India and *P. citri* has been consistently found associated with declining groundnut crops in India.

Among the *Xiphinema* species, the most common in the tropics, from West Africa to India and Sri Lanka, are *X. ifacolum,* which is highly pathogenic to rice and many vegetable crops, and *X. elongatum* which is a parasite of olive in Egypt and sugar cane in West Africa and India.

Methods of Control

Possible control measures vary with the paedoclimatic conditions, the socio-economic situation, the economics of the crop, the availability of chemicals and/or resistant cultivars, and the feasibility of some agricultural practices.

Crop rotations can always be adopted against species with narrow host ranges. However, plant sequence and time intervals between susceptible crops depend on the nematode species. For ectoparasites, with long life cycles and with a few generations each growing season, one-or two-year intervals might be appropriate. In the case of root-knot nematodes, or at least with species that lay eggs in a gelatinous matrix, at least three years between susceptible crops must elapse. But when cyst nematodes occur, even five years may not be sufficient to obtain economic results. High-value cash crops are often grown intensively by specialist growers and so it may be difficult to find a suitable plant sequence that will not alter the economic and cultural balance of a farm.

Resistant cultivars are without doubt the easiest and most convenient choice for the farmer. However, they are not always available, they may bring on to the market innovations not easily and quickly accepted by consumers, and they can develop resistant pathotypes or races of the parasite if used in monoculture for many years. Nematicidal treatments have been successfully used, but nematicides are usually expensive and may raise problems of environmental pollution and/or of accumulation of toxic residues in edible plant products. Methyl bromide is one of the most effective fumigants to control a broad range of problems (nematodes, fungi, insects, weeds). Being highly toxic to humans and highly volatile, it must be applied by skilled personnel and under a plastic tarp. Removal of tarping should be done two days after treatment and planting not earlier than one week after application. The high cost of the operation, lack of licensed operators and unavailability of the product, which requires proper storage facilities, are limiting factors for the use of this chemical.

A cheaper fumigant with predominant nematicidal effect is 1,3 dichloropropene (1,3-D). Its volatility is much lower compared with methyl bromide, therefore it does not require tarping. However, to be effective it needs to be injected into the soil by efficient machines which are not always available in developing countries and require accurate maintenance practices. The easiest and most effective way of applying 1,3-D is by injector gun, but this requires much labour. An interval of at least three weeks is necessary between application of this chemical and planting to avoid phytotoxicity.

When fumigants are applied, the soil must be carefully prepared by breaking plough pans and large soil aggregates and removing residues of the previous crop. Soil humidity and temperature must also be optimal to achieve satisfactory results.

Granular nematicides are more easily applied and safer for farmers compared with fumigants. However, they need high soil humidity to be effective. Those with systemic action are usually very effective and persistent.

Aldicarb, a carbamate, exerts its nematicidal activity for over six months, but it may cause problems of residues in edible plants. Therefore its use is allowed mainly on ornamental and industrial crops. Fenamiphos persists in the soil for about two months and if applied at rates of 15 to 20 kg a.i./ha gives good control even of root-knot and cyst nematodes without inducing accumulation of toxic residues in plants. It should be applied preferably two to three weeks before planting. Oxamyl is also a systemic nematicide but has shorter persistence (two weeks) and is less effective in tropical climates than carbofuran. Prophos performs its activity only by contact. Therefore it is effective against ectoparasitic nematodes and can control endoparasites only before they penetrate the roots. The limiting factor of all these chemicals is their high cost which makes them affordable only for cash crops. Even then, they are often not available on the market because hard currency is needed to import them, which is usually scarce in developing countries.

Two other pesticides that have mainly fungicidal activity have been used to control nematodes. To be effective dazomet needs moist soil and an interval of about six weeks between treatment and planting because of its high phytotoxicity. Metam-sodium has the same mode of action as dazomet, but has the advantage of being applicable through the irrigation system a few weeks before planting, and the disadvantage of having only modest nematicidal activity. Therefore, it needs very high rates of application to control root-knot or cyst nematodes, which are in most circumstances the only plant nematodes whose control gives economic benefit.

The hypothesis has been advanced that continuous application of nematicides might develop resistance to them in some nematodes. Up to now there is no evidence of resistance of nematodes to chemicals, and the cases in which low mortality of nematodes has been observed following nematicidal treatments have been interpreted as an enhanced degradation of the chemical itself resulting from the increased activity of microorganisms.

Attempts have been made to control plant-parasitic nematodes by soil amendments of by-products of some agricultural industries, such as neem, mustard, castor bean, groundnut or coconut oilcakes. In fact they reduce numbers of nematodes and female fecundity, but they are effective only when modest populations occur.

Plant extracts such as those from *Tagetes, Ruta, Cineraria or Pelargonium* have also been effective in killing plant-parasitic nematodes, but results refer mainly to *in vitro* or pot experiments and practical application of these extracts has yet to be profitable.

Soil steaming is often suggested as an environmentally clean measure to control plant pathogens. However, this time-consuming method can be applied only on small areas and, for the amount of energy (oil or electric power) required and the sophisticated equipment needed, is one of the most expensive nematode control practices. It is worth while only in greenhouses for production of out-of-season luxury foods. Besides, it does not always give uniform results and, if excessive temperatures are produced, reduces soil fertility and deteriorates soil structure.

Soil solarization is a recently developed technique in warm areas. It is a hydrothermal process that occurs in moist soil covered with transparent plastic film in summer months. During solarization, soil temperatures reach values (42° to 50°C) that are lethal to plant-parasitic nematodes, or increase to 36° to 40°C which, if prolonged, may be lethal or may affect life cycles and host-parasite relationships of many plant-parasitic nematodes by depleting their energy reserves. This makes them more vulnerable to biotic (antagonistic organisms) and abiotic stresses. However, temperatures of solarized soil are seldom lethal or sublethal at levels deeper than 30 cm. Therefore it is most practical with annual crops with shallow root systems. Its effect can be further improved if used in combination with reduced rates of fumigants or non-volatile nematicides.

Biological control may be feasible if antagonistic organisms are cheap to mass produce and apply. Easy colonization of soil, high virulence, long persistence, good shelf-life, low sensitivity towards agrochemicals, safety for operators and the environment and effective nematode control at acceptable costs for farmers are important factors. Unfortunately the fact that none of the organisms so far used for biological control of nematodes possesses all these characteristics and the lack of technology for producing commercial preparations of such bio-control agents are the main limitations to this practice.

In conclusion, different methods of nematode control could be combined in an integrated system, shifting from the concept of control to the concept of management which is a procedure directed at reducing and maintaining the numbers of plant-parasitic nematodes at non-injurious levels. This is particularly important in developing countries, where the lack of a commercial system and of a market for wealthy customers often means that high production costs are not justified. However, demographic increases require improvements in production from sustainable agriculture.

Regional Climates and Vegetation

A forester faced with the practical problems of choosing what species to plant on a given site, or of selecting sites and species for the production of a particular type of timber, needs a detailed knowledge of the local climates and plant associations of his territory. This will direct him to other areas on which he might be able to draw for additional species. He will, of course, have made himself familiar with all available relevant experience whether published or to be found in the files of the various organizations concerned with natural history, land use, biological, geological, climatological and hydrological research in his country. He will also have studied the topography, soils and vegetation, both natural and artificial, of the tract in question and of other parts of the country having similar conditions.

If he believes that his purpose may be served best by the trial of exotic species, he will obviously first consider making introductions from homoclimes or climatic analogues. He will turn to zones of similar and similarly distributed rainfall and temperatures with similar latitudes and topography, and seek detailed information about the distribution and growth there of useful species, and the types of soil on which they grow.

These zones will be indicated somewhat vaguely on climatological maps of the world, and more detail about them will be available in local regional maps and in tables of climatological data published in the regions. Then the forester can generally turn to a local flora of the region to ascertain what grows there, and often he will be able to follow up the species which arouse his interest in monographs or other papers dealing with them.

If he is fortunate, he will find works such as *Applied Silviculture in the United States* (Westveld, 2nd. ed. 1949) and *The Silviculture of Indian Trees* (Troup, 1921), which will give him climatic, ecological and silvicultural information about the trees of considerable areas. Records of trials of species in different parts of the world such as are to be found in *Exotic Forest Trees in the British Empire* (Troup, 1932) will also be useful. If he has reason to think that some members of the genus *Eucalyptus* might be of value to him, he can turn to the FAO publication *Eucalypts for Planting* (Metro, 1954) the various reports of the FAO Eucalyptus Study Tour (1952) and others of the 300 or so publications listed in the bibliography of that book, and the *Forest Seed Directory* (FAO 1956), which lists 118 eucalypt species classified according to their suitability to grow in any of 18-climatic zones. Other Australian trees besides the eucalypts are included in *A Reconnaissance of the Forest Trees of Australia from the point of view of their cultivation in South Africa* (Robertson, 1926).

There is, of course, nothing new about this. Horticulturists and foresters have for many years looked to countries with climates similar to their own, visited them and brought back seed of species which they thought might make promising additions to their local floras. The only difference now is that very much more information is available, and that reliable seed can be obtained from the zones selected without difficulty or resort to smuggling as was sometimes necessary in the past.

As an example, it may be noted that in 1823 Douglas was sent to North America by the Royal Horticultural Society of London because it was realized that the Pacific Coast region was the home of a particularly rich flora growing in a climate comparable to some parts of the British Isles. This region from Oregon northwards through Washington and British Columbia to southeast Alaska and bounded on the east by the Coastal Ranges contains a far wider range of climate than does Britain, but the temperature regime at the mouth of the Columbia River around latitude 46° N., with an average in summer of 54° F. (13° C.) and an average for the year of 50° F. (11° C.), is similar to that of the English south coast, latitude about 50.5° N.; and Sitka in Alaska, latitude 57° N., closely matches the northwest coast of Scotland. The rainfall on the Pacific coast is of the winter type with midsummer the driest period, whereas in Britain spring is relatively dry. Further north up the coast the proportion of rain falling in the summer is greater, until in southeast Alaska it is close to the 22 percent commonly found in Britain. Inland, the climate becomes rapidly continental and summer temperatures are higher than in England.

The trees of the region important to forestry in Britain and parts of western Europe have been found to be Sitka spruce, Douglas fir, *Pinus contorta* Dougl., *Abies grandis* Indl., *Tsuga heterophylla* Sarg. and *Thuya plicata* Don. All penetrate into climates of much lower winter temperature than are found in Britain, and Douglas fir and the *Abies* than those of Britain, in particular grow best in warmer summer temperatures averaging 52° to 56° F. (11.1°-13.3°C.) though there is no reason to suppose that they are limited by summer temperature (Wood, 1955). In their native habitat, neither Douglas fir nor Sitka spruce appear to form part of climax forest communities, but are pioneer species giving way to *Tsuga, Thuya* and *Abies*. Sitka spruce thrives at low elevations in the British Columbian region with mean annual temperatures varying from 44° to 49° F. (6.7°- 9.4°C.), summer mean of 55° to 68°F. (12.8°-20°C.), winter mean of 30° to 38° F. (1.1°-3.3° C.) and a rainfall of 40 inches (1,020 mm.) a year upwards, conditions which can easily be matched in Britain.

It is now possible to buy from government forest departments and from merchants seed certified as collected from known climatic and elevational zones in this region as from several other regions of the world. The *Forest Seed Directory* mentioned above contains the forms of agreed international certificates of origin and quality, and the sources from which a large number of species can be obtained. One firm divides its collection area on the Pacific coast from Alaska to California into 25 climatic regions, subdivided into 127 zones, each of which is further split up into 500 foot (150 m.) elevation bands. It provides tables which give the temperatures (average annual, average summer, absolute maximum and absolute minimum), precipitation (annual and summer), and the number of frost-free days in the year, for the most representative weather station in each zone.

Another country which was drawn on early for plantation species was Australia. In 1843, several species of eucalypt were tried in the Nilgiri Hills of India, and later *Eucalyptus globulus* Labill. solved a serious fuel shortage there. In 1870, *Acacia mollissima* Willd was introduced into Natal and eastern Transvaal in South Africa and by 1952 a great tanbark industry depending on some 640,000 acres (259,000 ha.) of black wattle plantations had been built up. In 1876, *Eucalyptus saligna* Sm. was first planted for railway fuel in South Africa, and now there are some 200,000 acres (80,000 ha.) of it in eastern and northern Transvaal and on the Zululand coast. Early in the present century, the *Companhia Paulista de Estradas de Ferro*, the State railway of Sao Paulo, Brazil, imported 143 species of eucalypts and now has some 30,000 acres (12,000 ha.) of plantations, mainly of *E. camaldulensis* Dahn., *E. citriodora* Hook and *E. saligna*. The Mediterranean countries, Spain, Portugal, France, Italy and Greece, have all found *E. globulus* and *E. camaldulensis* valuable additions to their plantation species. In California, 80 percent of the considerable eucalypt plantations are of *E. globulus*, and 15 percent of *E. camaldulensis*. Peru, Chile and Uruguay all make considerable use of *E. globulus*, while in Turkey, Cyprus, Palestine, Jordan, Tripolitania and Morocco the most popular eucalypt is *E. Camaldulensis*. Many other species are found to be suitable to particular parts of these and many other countries.

It is, therefore, worth considering the climate and vegetation of the Australian continent. Australia has a central desert zone covering nearly 40 percent of its whole area, and the rest can be divided very roughly into three main climatic zones:

1. A tropical northern zone of summer rainfall, north of about 20° S. on the west coast and about 14° S. on the east coast. The mean annual rainfall varies from about 20 inches (510 mm.) in the south of the zone to 60) inches (1,525 mm.) and more on the north and east coasts, with some areas receiving 120 inches

(3,050 mm.). The mean annual temperature is in the neighborhood of 80° F. (26.4° C.), the interior being warmer than the coastal areas and the west coast warmer than the east.

2. A zone of moderate winter rainfall and dry summers in the southern parts of South Australia, Western Australia, the greater part of Victoria, and the south of New South Wales. Mean annual rainfall varies from about 10 inches (250 mm.) to about 30 inches (760 mm.) with isolated areas of up to 40 inches (1,020 mm.) on the south coast. Summer temperatures are around 70° F. (21.1° C.) and winter temperatures around 35° F. (1.7° C.)
3. A zone of more uniformly distributed rainfall on the east coast and the southeastern part of the continent with annual rainfall generally between 20 inches (510 mm.) and 60 inches (1,526 mm.). This zone includes the mountains of New South Wales, and there and on the east coast the rainfall is from 40 to 60 inches (1,020-1,526 mm.) a year, with isolated areas of 120 inches (3,050 mm.) on the coast of northern Queensland, where the summer is the wetter season. In the south (Victoria), there is a tendency towards a maximum rainfall in winter. In New South Wales, the even distribution is more noticeable, though in the north and on the east coast more rain falls in summer, and on the south coast more in winter. There are very considerable variations in temperature regimes within the zone, because of the topography.

De Beuzville (1943) considered that the whole of the east coast of Australia has a climate predominantly similar to that of Argentine, Brazil and Mexico and certain parts of South Africa, and also, in its wetter parts, in Madagascar. The north coast and its hinterland has a climate similar to those of parts of Madagascar, India and Mozambique. On the west coast and inland from it there are stations with climates similar to some found in South Africa, Algeria, Sudan, Mexico and Arizona. The climate in the south southwest is more like that of California and South Africa, and in small areas like that of Spain. The Australian Alps and the Dividing Range have cool slopes where the climate approximates to that of France and Italy. The dry centre of Australia has a climate similar to that of Egypt, Arizona, Israel and parts of the Californian desert. Robertson (1926) found that, while coastal temperatures, latitude for latitude, were very similar in Australia and South Africa, the interior of Australia was warmer than that of South Africa because of the greater elevation of the latter.

The forest formations of Australia may be roughly divided into Rain Forest (tropical, sub-tropical, and temperate), Moist Sclerophyllous Forest, pry Sclerophyllous Forest, and Open Woodland, all of which merge into each other.

The rain forests occur only in isolated relict patches. In the tropical zone their flora is Malayan without predominant species, and the natural succession is very confused. From the Hastings River, about 30° S., in New South Wales to the north of Queensland, *Araucaria cunninghammii* Ait is found on the coast and up to 100 miles (160 km.) inland with an annual rainfall of 50 to 70 inches (1,270-1,780 mm.), falling mainly in summer. *Araucaria bidwillii* Hock, and *Agathis robusta* are found in southeast Queensland in areas with 50-60 inches (1,270-1,525 mm.) summer rain, with summer temperatures of 75° to 85° F. (24°-29° C.) and winter temperatures of 50° to 60° F. (10°-16° C.). What are apparently secondary forest formation after fires include *Tristania conferta* R. Br. and *Syncarpia laurifolia* Ten. mixed with eucalypts such as *Eucalyptus saligna*.

The sclerophyllous formations are mainly composed of eucalypts and acacias. Closed stands of Eucalyptus are not usually found where the rainfall is less than 30 inches (760 mm.) a year, below which limit open and scrubby vegetation occurs. Many of the eucalypt species are confined to particular zones of rainfall amount and distribution, but some are very adaptable. One of the most widely distributed is *E. camaldulensis* which Boosman (1950) states can put up with 5 to 10 inches (130-250 mm.) of rain a year, is often found in zones of 22 to 28 inches (560-710 mm.), and exceptionally with 40 to 45 inches (1,020-1,140 mm.) though 35 inches (890 mm.) is regarded as its normal maximum. Actually its range is determined more by soil moisture conditions than by precipitation and it is characteristic of alluvial flats, often along river banks where water is available by inundation or percolation. It likes warmth and extends into some of the hottest parts of the Northern Territory. It is not found in the highlands of New South Wales in which State de Beuzville (1943) gives a mean temperature of the coldest month of 46° F. (7.8° C.) as the lower thermal limit of its range. It must be remembered that the form, as well as the stand density, of eucalypt species growing in areas of their minimum rainfall is very different from what it is under optimum rainfall.

Moist sclerophyllous forest is found mainly in the southeastern coastal strip of New South Wales and Victoria, and in the southwest corner of Western Australia, in the zones of either winter or fairly uniform rainfall. In Western Australia, *Eucalyptus diversi-colour* F. V. M. is found in stands of a type approaching temperate rainforest in a climate of 40 to 60 inches (1,020-1,525 mm.) mean annual rainfall and mean annual temperatures of 62° to 65° F. (16.7°-18.6° C.) In Victoria and New South Wales, *Eucalyptus globulus* grows in the higher and cooler sites with 35 to 65 inches (890-1,650 mm.) of rain and mean annual temperatures of 44° to 49° F. (6.7°-9.4° C.) sometimes mixed with *Acacia melanoxylon* R. Br. and with an understorey of *Acacia mollissima* Wild. or *A. dealbata* Link. In the wetter range of these sites in Victoria, stands of *Eucalyptus regnas* F. v. M. are found.

Eucatyptus gomphocephala A. Dc. with an undergrowth of *Acacia cyanophylla* Lindl. is found growing on calcareous sands in the southwest of Western Australia in an annual rainfall of 30 to 40 inches (760-1,020 mm.) mainly in winter, with an annual mean temperature of 60° to 64° F. (15.6°-17.8° C.) and a summer temperature of 67° to 72° F. (19.4°-22.2° C.). There are either no frosts or only occasional light ones in places. Both species have proved valuable in the afforestation of coastal sand in South Africa, and more recently have shown considerable promise when used in combination on calcareous sand dunes in Morocco (FAO, 1954).

In the dry sclerophyllous forests there is a very great variety of eucalypt species, not many of which attain timber dimensions in the lower rainfall areas. In New South Wales and Queensland, *Eucalypts crebra* F. V. M. grows with a rainfall of 20 to 26 inches (510-660 mm.) a year mainly in summer in areas of high temperature, often mixed with *Callitris glauca* R. Br. *Eucalypts sideroxylon* A. Cunn. is found in poor sites in New South Wales, Victoria and Queensland with about 20 inches (510 mm.) of rain a year. There are also species which are characteristic of moister forests but nevertheless can grow with as little as 22 to 25 inches (560-635 mm.) a year, such as *Eucalypts marginata* Sm. in Western Australia, and *E. oblique* l'Herit. in South Australia, Victoria and New South Wales. Where there is moisture enough to permit of under this is often *Acacia pycnantha* or *A. mollissima* in the east, and *A. cyanophylla* or *A. cyclops* Cunn. in the west.

These formations merge into open woodland, particularly in western Queensland and New South Wales. *Eucalypts salmonophloia* F. V. M. grow in Western Australia in the zone of 9 to 12 inches (220-305 mm.) of rain, mainly from June to September. 'Mallee' formations consisting of a great variety of small, often bush-like, eucalypts attaining up to 20 feet (6 m.) in height with stems of 3 to 6 inches (7.6-15.2 cm.) in diameter, cover considerable areas of low rainfall.

South Africa is a good example of a country with inadequate natural forests and slow-growing indigenous species. The need to supplement the natural flora was recognized early. Conifers and hardwoods were imported from northern Europe from 1665, but did not thrive. Later some Mediterranean species, notably *Pinus pinea* L., which became naturalized in the Cape Peninsula, and *Pinus pinaster* Sol., which has been widely planted and does best in the zone with winter or all-the-year-round rainfall, were introduced. By the middle of the nineteenth century *P. radiata* had been brought from California and had proved to be very adaptable in climates similar to those favoured by *P. pinaster*, though more particular about soil. Later in the century, *P. longifolia* Roxb. was introduced from the outer ranges of the Himalayas in India and proved useful in the cool and warm temperate summer rainfall areas.

A great step forward was made about 1905 when Hutchins decided that Mexico, a plateau country like the Transvaal, lying in approximately similar latitudes, having a summer rainfall and a rich flora, should be a promising source of exotics. He tried a number of species including *Pinus patula* Seem. of which there is now a bigger acreage in South Africa that of any other softwood, particularly, of course, in the summer rainfall areas. Hutchins also recommended *P. caribaea* Morelet from the southern United States of America for the sub-tropical zone, and it has shown exceptionally vigorous growth in both the temperate and sub-tropical zones with winter or all-the-year-round rainfall. It is being planted on a large scale for paper pulp in Zululand. By 1947, some nervousness was felt about the dangers of disease in large pure blocks of these two species and an officer was sent to America to study the pines in their natural habitats and to find additional species if possible (Loock, 1950). He was particularly interested in *Pinus pseudostrobus* Lind. which had originally been tried under the name of *P. teocote* Schl. v Cham. with success, but the seed of which could not be obtained. He was able to collect 350 pounds (159 kg.) of seed in Mexico. He was also impressed by *Pinus hondurensia* Loock in British Honduras, only lately distinguished from *P. caribaea.* This illustrates the importance of botanical identification and of securing seed of the most suitable species and varieties for use in particular localities. Where reliable local agencies for the supply of seed do not exist it may be necessary to go and collect exactly what is wanted.

These conifers and others from California, Central America, India, the Mediterranean region, and Mexico now cover more than 600,000 acres (243,000 ha.) of South Africa, alongside some 640,000 acres (259,000 ha.) of acacias and 450,000 acres (186,000 ha.) of eucalypts from Australia, an example of what can be done by selection and trial of species from climates similar to those in which they are required to grow.

Reference is made earlier, in the study of which this article is an exerpt, to the agro-climatic analogues established at the American Institute of Crop Ecology by Nuttonson (1937-53). Up to the latter year analogues had been established between North America and Ukraine, Poland, Czechoslovakia, Yugoslavia, Greece Albania, China, Germany, Finland Sweden, Norway, Siberia, Japan, and the Ryukyu Islands. Each of these studies gives a general and comparative geography of the country covered and a brief account of its climate. There is also a varying amount of information about the natural resources of the country, its soils, major plant species, land-use, agricultural and forestry practices, etc. Relief maps show the location of the meteorological stations within the country studied and the regions of any North American climatic counterparts found. Charts give full meterological data for each station and for the counterparts.

The purpose of this series of studies is "to formulate an agronomic and horticultural approach to ecology and to promote research along the lines of crop ecology as it affects plant adaptation, plant introduction, and the exchange of varietal plant material among the various agricultural areas of the worlds". Its value is not, however, confined to agriculturists. For instance, the paper on Japan (1951) points out:

"Over one half of the total land area of the country is covered with luxuriant forests of a great diversify of broadleaved and coniferous species and of dense undergrowth. These forests are found largely in the mountainous regions, as more than three-fourths of the country's land consists of rough and rugged mountains and hills topographically unsuitable for cultivation of farm crops."

The climate of Japan is largely determined by the prevalence of the monsoons. There is an abundance of precipitation throughout the year with a predominantly summer-maximum distribution pattern, high relative humidity, and high summer temperatures, except in the higher latitudes and altitudes. Nuttonson mapped and described eight climatic regions, two of which he subdivided into three and two districts respectively, and provided graphs of average monthly temperatures, monthly rainfall, and frostless seasons from 16 weather stations. Unfortunately, because the study is mainly concerned with agriculture, not a great deal of information is given about the climates in the mountain forests, where the trees of greatest interest to other countries grow. Data for two high level stations in central Honshu are:

Latitude	Elevation		Mean annual rainfall		Mean annual temperature		Mean Jan. temperature		Mean August temperature	
	ft.	m.	inch.	mm.	F.	C.	F.	C.	F.	C.
35° 28'	3 687	1 124	73.9	1 877	45.3°	7.4°	24.8°	4 0°	66.0°	18.9°
36° 25'	4 514	1 376	92.1	2 339	41.7°	5.4°	19.9°	6.7	64.0°	17.8°

They are probably typical of the natural habitat of two of the most important conifers, *Cryptomeria japonica* Den. and *Chamaecyparis obtusa* Sieb. About one-third of the total forest area of Japan consists of plantations and they are being extended at a rate of about 720,000 acres (292,000 ha.) a year. In the best sites in the moister inter-mountain valleys and on the lower slopes *Cryptomeria* is planted, on the middle and upper slopes *Chamaecyparis*, and on the ridges from about 6,000 to 7,000 feet (1,830-2,130 m.) elevation, *Pinus densiflora* Sieb. According to Turner (1937), *Larix leptolepia* Murr. is the species most used for planting in Hokkaido. It is also native to the volcanic mountain soils between altitudes of 4,000 and 6,000 feet (1,2201,830 m.) elevation.

Nuttonson divided the natural forests into three climatic types, each of which descends in general to lower altitudes in the northern latitudes and ascends in the southern. This classification agrees fairly closely with one drawn up in 1922 which is quoted by Ohmasa (1956).

1. *Evergreen hardwood forest*, which contains some deciduous hardwood trees and some conifers. It lies mainly south of latitude 36° N. and covers Kyushu from sea level up to about 2,800 feet (850 m.) elevation, Shikoku to 2,450 feet (750 m.) and southern Honshu to about 1,800 feet (550 m.). The mean annual air temperature is 55°-70° F. (13°-21° C.) and there is a well-distributed rainfall with a summer maximum. The more important species are *Quercus gilva*, other

Quercus spp. and *Cinnamomum camphora*. *Pinus thunbergii* Parl. is found on the sea coast.

2. *Deciduous hardwood* forest, which includes mixtures and pure stands of economically important conifers, is found to the north of the evergreen hardwood forest and at elevations above it, reaching latitude 43.5° N. at sea level in the middle portion of Hokkaido. The formation covers the highest elevations found in Kyushu, but only land up to 6,500 feet (2,000 m.) elevation in Shikoku, to 3,600 feet (1,100 m.) in northern Honshu, and to 1,500 feet (450 m.) in southern Hokkaido. The annual mean air temperature of the zone is from 43°-55° F. (6°-13° C.) Valuable hardwoods include *Fagus crenata, Quercus crispula, Aesculus turbidata* and *Zelkova serrata*. The principal conifers are *Cryptomeria japonica* Den. *Chamaecyparis* spp., *Larix kaempferi* Lamb. and *Pinus densiflora* Sieb.
3. *Coniferous forest* covers the colder northern and more elevated tracts. It does not occur in Kyushu, and in Shikoku only in the small area above 6,500 feet (2,000 m.) elevation. In central Honshu, it is found between about 6,000 and 8,900 feet (1,800-2,700 m.) In southern Hokkaido, the formation covers most of the land above 1,000 feet (305 m.) elevation, and north of latitude 43.5° N. descends to sea level. The annual mean air temperature is 43° F. (6° C.) and below. *Picea jezoensis* Sieb. and *Abies sachalinensis* Mas. are the principal conifers, and are sometimes mixed with *Quercus crispula Tilia japonica, Betula tauschii*, etc. *Abies veitchii* Lindl. occurs in the colder parts of the zone.

High relative humidity is a feature of most of Japan, and, while similar thermal conditions to those found in different parts of its wide range of latitudes can be found in North America, it is not so easy to match the distribution of rainfall. However, some homoclimes have been found in the Maritime Provinces of Canada, and many in the northern, central and southern regions of the eastern United States.

Nuttonson, 1947, 1951, has pointed out that in many parts of North America the relatively severe winters compared with Japan do not permit the formulation of year-round thermal analogues in a number of areas in which temperatures in the growing season (April-September) should make possible the harvesting of spring planted crops, and many spring-crop analogues are suggested. Similarly there are winter-crop analogues based on October to April temperatures. This points the difference between agriculture and forestry, in which crops are on the ground throughout the year and year-round temperatures are of more importance.

Protection of Soil and Watersheds by Vegetation

It is well to keep in mind that the primary objectives of soil conservation and watershed management are to stabilize and maintain the natural resources to the greatest extent economically feasible. On wild lands this can best be done by maintaining an abundant and vigorous plant cover.

By maintaining a desirable porosity and permeability the soil absorbs rainfall readily, hence erosion is greatly minimized. The roots also function to maintain permeability. The litter furnishes food for microorganisms which help maintain soil permeability, and the organic matter gives the soil properties to resist compaction. Another important function is the sheltering effect of the soil from direct insolation and desiccating winds, and from the full force of rainstorms. On well-vegetated lands a goodly portion of the rainfall is intercepted and released gently to the soil. On the other hand, faulty logging, excessive grazing and wild fires are among the most common causes of upsetting plant-soil relations.

Another effect of vegetation on watershed management is its capacity to withdraw water from the soil and thus, in a measure, reduce water yield. Kittredge has estimated annual evapo-transpiration loss from southeastern pine forests at 30 inches (75 cm.) and oak-hickory forests at 20 to 30 inches (50 to 75 cm.). Hoover found that a hardwood forest removed from 17 to 22 inches (43 to 66 cm.) of water annually. Since the anatomy of leaves of the various species composing the plant mantles are different, transpiration 1088 would also be different. To offset this seeming loss by transpiration is the fact that the soil mass is again receptive to water absorption as more rain falls, thus minimizing run-off and erosion.

Closely related to water loss is the character, volume and distribution of the root systems. Grasses have numerous fibrous roots which are localized in the upper few inches of the soil; trees have deep woody roots with few surface rootless; shrubs also have woody roots but they are less coarse than tree roots and are anchored at depths intermediate to those of grasses and trees. One plant community, therefore, may withdraw moisture from the soil over a much longer period than another composed of different species. On inferior steep slopes brush species, in contrast to grasses and trees, will send down "stringer" roots in rock crevices which will maintain these frugal plants indefinitely. Bleak watershed units should be left undisturbed as no vegetation will serve better than the well-adapted natural cover which they already support. It is, therefore, necessary for the land manager to learn as much as possible about rates of water 1088 and root system characteristics of the dominant trees and shrubs of his land. Protection of watersheds against uncontrolled fires is of utmost importance because of the danger of serious damage to the soil. The remaining plants may be deprived of needed water, and the run-off carries with it plant nutrients, ashes, and surface soil, thus lowering the quality of the site. Further soil depletion accrues during dry weather from sheet erosion.

The time required for a watershed to recover from a severe burn varies according to climate, type of soil, steepness of slope, length of time the soil remains bare, and several other such considerations. Recovery may take several decades or even centuries where much soil has been lost. If left to nature, recovery will pass through the entire sequel of successions and reach the tree stage or climax when the soil is again productive. Seeding and planting, though costly, will greatly speed up the time of recovery.

Whether the amount of usable water is increased or diminished by burning and thinning of the stand remains a moot point. More snow or rain will reach the soil on burns, but surface run off is correspondingly greater.

Suggestions for Good Land Management Practices

Nearly all operators and administrators have encountered from time to time unexpected if not baffling problems in maintaining the desired returns from their lands. Some of the areas may be in need of reforestation, other parts may have a thinned out forage cover, still others may have become brushy. The suggestions which follow may be helpful in meeting some of the problems involved.

Seeding and Planting: *Reforestation*. There is urgent need to re-establish trees on forest lands where the timber has been logged or burned off. After fire or logging watershed values are lowered, and a new tree stand should be started to stabilize the soil and to put the land back into commercial use. This is expensive and requires skill.

Reforestation by natural means is relied on where adequate trees of desirable species remain. On heavily burned or improperly logged areas, on the other hand, seeding or planting of trees is the only means available for restocking the land.

Range Seeding: Prompt re-establishment of a forage cover on cleared grazing land is important for three reasons:

(a) to protect the soil from excessive erosion;

(b) to curtail establishment of brush seedlings;

(c) to provide feed for livestock.

In order to protect the soil, the land should be revegetated the first year after clearing. By careful selection of site and species, and by proper planting and management, earlier and more palatable forage may be obtained than the lands produced when in their virgin state.

Outstanding Features of Range Management: Attempts seeding of cleared brush or forest lands are doomed to failure if grazing is allowed before the resulting seedlings have developed a strong root system. The fact must not be overlooked that these woody species are hardy and vigorous, thus indicating how well adapted they are. The species that are to be introduced, therefore, must be given all cultural advantages possible over the potential reinvasion of the hardy woody plants. In the western United States, areas seeded to perennial grasses should have at least one full year, and in many instances two years, of total protection before grazing is allowed. Rodents should be poisoned where numerous, and, if possible, heavy invasions of weeds should be mowed in early summer or sprayed with a selective herbicide.

Satisfactory distribution of livestock over the range is of great importance in order to avoid spotty overgrazing and undergrazing. On cattle range this can be dope by salting away from the water, development of watering places remote from meadows and other favourite units, and by fencing to prevent drifting and better control of the animals. On open sheep range the herder should so direct the flock that the grazing pressure is properly distributed. One of the cardinal principles of range management is that of determining and applying the proper season of use. The opening date for grazing should be delayed until growth is far enough along to afford a "good bite," and until the soil has become firm. Generally the grass leaves should be 3 to 5 inches (8 to 13 cm.) in length and all the free gravitational water should be drained away. A close relation exists between seasons of use and numbers of livestock grazed, for grazing cape capacity is determined in terms of total animal months use. Range that extends into different life zones should be grazed on a seasonal zonal basis. The earliest units should be grazed first, followed by the next higher elevational zone.

Harvesting of the forage crop should be conservative. Too close grazing ultimately leads to reduced forage yield and loss of soil fertility. Appraisal of range utilization, once a proper standard of use has been established for a given area, can be made by estimation, by measurement, or by a combination of these.

Most perennial ranges are properly stocked when about 60 percent of the year's growth is left on the ground at the end of the grazing season. This would be roughly equivalent to a stubble height of 4 to 6 inches (10 to 13 cm.) for the tall grass type composed of plants like the wheatgrasses, fescues, and bromes, and 1 ½ to 2 inches (4 to 6 cm.) for the short grass type, as for blue grama and buffalo grass.

Generally range utilization is uneven because of differences in preference of livestock for the various species. For practical reasons, the degree of utilization is based on so-called "key" species, or some two or three kinds of plants that provide most of the forage. When the key plants are properly utilized-as at the end of the grazing season-the range as a whole is also assumed to be properly grazed.

One complication in blooper stocking is that forage production may fluctuate widely from year to year because of differences in precipitation and temperature. Accordingly, stocking rate should be based on production between the average and the poorest years. The question of stocking must be settled for each individual ranch or administrative unit as local conditions and experience would indicate.

Indian Floriculture Industry Present

Status & Scope: After liberalization, the Govt. of India identified floriculture as a sunrise industry and accorded it 100 percent export oriented status. Owing to steady increase in demand of flower floriculture has become one of the important Commercial trades in Agriculture. Hence commercial floriculture has emerged as hi-tech activity-taking place under controlled climatic conditions inside greenhouse. Floriculture products mainly consist of cut flowers, pot plants, cut foilage, seeds bulbs, tubers, rooted cuttings and dried flowers or leaves. The important floricultural crops in the international cut flower trade are rose, carnation, chrysanthemum, gargera, gladious, gypsophila, liastris, nerine, orchids, archilea, anthuriu, tulip, lilies.

Present status and growing trade is still in infancy. Floriculture in India, is being viewed as a high growth Industry. Commercial floriculture is becoming important from the export angle. The liberalization of industrial and trade policies paved the way for development of export oriented production of cut flowers. The new seed policy had already made it feasible to import planting material of international varieties.

The government of India offers tax benefits to new export oriented floriculture companies in the form of income-tax holidays and exemption from certain import duties. Agricultural and Processed Food Products Export Development Authority (APEDA), responsible for export promotion and development of floriculture in India, grants subsidies for establishing cold storage, Pre-cooling units, refrigerated vans and green houses, and air freight subsidy to exports. It has been found that commercial floriculture has higher potential per unit area than most of the field crops and is therefore a lucrative business.

According to a report of the APEDA, the total area under flower crops was estimated around 34,000 hectares, which included 24,000 hectares under traditional flowers such as marigold, jasmine, aster, rose, chrysanthemum, tuberose and 10,000 hectares under modern flowers like cornation, rose, gerbera, gladiolous, anthurium.

Returns from floricultural products were estimated at Rs.205 Crores, which included Rs.105 Crores from traditional and Rs. 100 Crores from modern flowers. The data collected by National Horticultural Board (NHB) (1994-95) show that estimated area and production were 53,000 hectares and 4 lakh tonnes. In 1995-96, there were mainly 230 export-oriented units in India. More than 50% of the floriculture units are based in South zone mainly in Karnataka, Andhra Pradesh, Tamil Nadu. Also West Bengal, Maharashtra, Rajasthan have large areas under floriculture. The domestic flower production goes on increasing annually. Technical collaborations with foreign companies have been approved for India, in order to increase total share in the floriculture world trade.

Conductive Conditions: India is endowed with diverse agro-climatic conditions like good quality soils, suitable climate, abundant water supply, low labour cost, proximity to market in Japan, Russia, South-East Asia, Middle-East Countries. Subsidy on airfreight for export of cut flowers and tissue-cultured plants is allowed by the Government. Freight rates are Rs.10 per kg for export to Europe and Rs. 6 per Kg for export to West Asia, South-East Asia whichever is less.

Import duties have been reduced on cut flowers, flower seeds, tissue-cultured plants, etc. Setting up of walk in type cold storage has been allowed at the International airports for storage of export produce. Direct subsidy upto 50 percent of the Pre-cooling and cold storage units is available, as well as subsidy for using improved packaging material is given by APEDA.

Eleven-model floriculture centre units and two large centres, 20 tissue culture units have been established by Ministry of Agriculture. Refinance assistance is available from NABARD to a number of hi-tech units at reasonable interest rate.

Export Constraints: In spite of an abundant and varied production base, India's export of floricultural product is not encouraging. The low performance is attributed to many constraints like non-availability of air space in major airlines, since most of the airline operators prefer heavy consignments. The existing number of flights during the peak seasons is not sufficient for export purpose.

Exporters for infra-structural problems like bad interior road, inadequate refrigerated transport and storage facilities. Lack of professional backup of delivery and supporting companies, which resort into high cost of technology for Indian entrepreneurs. Tedious Phyto-sanitary certification and an unorganized domestic market. In order to overcome these problems, attention must be focussed on: Reduction in import duty on planting material and equipment.

Airfreight should be reduced to a reasonable level. Sufficient cargo space may be provided in airlines. Establishment of model nurseries far supplying genuine planting material.

Cooperative florist organizations should be established at regional level. Training centres for diploma course on the pattern of ITI for training the personnel in floriculture should be set up.

Exporters should plan and monitor effective quality control measures right from production to post harvesting, storage, and transportation. An analysis of strengths, weaknesses, opportunities of the floricultural industry shows that India has immense potential for export of floricultural products.

Though the global floriculture industry is growing comparatively at a faster pace than in India, still a scope exists to bridge the demand and supply gap.

Rose Cultivars: Rose is the most popular flower the entire world over from times immemorial. Originally its fragrance attracted man's attention but later its colour, shape, size, recurrent flowering and vase life, etc. became the equally important attributes. To the naturally existing varieties in different agro-climatic localities, the rose enthusiasts added much more creations by utilizing the tool of hybridization for new and desired characters to incorporate maximum good qualities in a single variety. In the recent years another important pathway called genetic engineering is also being practiced to improve the qualities of available plant varieties and plant species. Although much success has been achieved in this direction so far efforts are continuously being done to develop still better varieties with the result that the number of varieties grown at any time indifferent agro-climatic situations is so enormous that any new rose enthusiast is usually puzzled to make a list of his choice. It is, therefore, convenient to know the major groups of rose cultivars commonly grown and which have persisted for several years in nature also.

Roses, thus, can be classified in different ways *e.g.* on the basis of purpose of growing *viz.* Garden roses, cut-flower roses, landscape roses, exhibition roses, fragrant roses, etc. on the basis of growth habits *viz.* Tall, medium, dwarf, climbers, ramblers, etc. or on the basis of botanical characteristics *viz.* Species roses. For practical purposes, however, certain criteria selected partly from growth habits and partly from botanical characteristics are the most suited for identifying the rose cultivars. Accordingly, the cultivated or popularly grown roses are usually and broadly grouped as follows:-

1. *Large flowered or Hybrid Tea (H.T.) Roses:* Roses in this group are thin to medium thick bushes with height varying from about 75 to over 200 cm. Their name, Hybrid Tea Roses, is derived from the old naturally occurring roses giving fragrance resembling to that of a freshly opened tea chest and which were used in the hybridization programmes to develop the modern roses. The plants are usually vigorously growing, perpetually flowering with large (6-10 cm. wide) blooms having 15-20 to 70 –80 petals with attractive shapes and colours. Although most of them have no or very little fragrance a few emit strong scents characterised as fruity, spicy, sweet, etc. The flower stems are medium to long (30 to 100 cm) and in most cultivars every stem bears only a single blook. This is the most widely grown type of roses for home or public gardens, exhibition purposes and also for the cut flower trade. On the basis of plant growth habit, colour and shape of the bloom and fragrance etc. there are several hundred varieties or cultivars in this group suited to different agro-climatic conditions.

2. *Clusters flowered or Floribunda Roses:* These are medium to very thick bushes–with height ranging from 50 to 120 cm. The blooms are of medium size having 6-8 to 20-25 petals, every stem often bearing 4 to 6 or even much more blooms.

The stems are small to medium in length. Since abundant blooms are commonly produced in every flush, the plants exhibit excellent beauty of nature. Several good varieties with varying colour shades are available in this group also. These are most suitable for planting along the roads in beds and for landscape designing as well.

3. *Miniature Roses:* As the name suggests, these roses can be well described as miniature floribundas. The plant height is about 30 to 50 cm, with tiny delicate leaves and flowers. When in full bloom the plants get fully covered with small attractive colourful flowers in thick clusters. These roses also make excellent pathway borders and individual potted plants.
4. *Climbing Roses:* These are mostly naturally growing wild roses with thick vine-like growth habit. The blooms are of medium size and they are usually produced in clusters. The blooming period is also short and seasoned as compared to that of the other three groups described earlier. Even though these roses are quite popular in western countries, they are very rarely grown in India, perhaps due to the fact that cultivars suitable for the agro-climatic conditions existing in the country are not available.

Besides the four groups of roses described above, the wild roses like 'Polyanthas', 'Ramblers', etc. are also seen in some gardens but the blooming period being short and seasonal and due to shy blooming nature these roses are not very popular.

Although hundreds of excellent rose varieties have been developed so far and are being grown popularly, the rose hybridizes are continuously trying to evolve still better cultivars suitable for different purposes and different purposes and different agro-climatic situations and several new varieties are released every year in different parts of the world.

Another important point to be noted in this respect is that howsoever excellent a variety may be, it may not be so under all circumstances. Climate and Soil of a locality profoundly affects the performance of a variety.

Annuals

Annuals are an important group of flowering plants widely used for garden decoration, cut flowers and pot plants. Few types, however, are grown for attractive foliage. They have large number of species and many varieties which not only show variations in height and growth habit but also in the shape, size and colour of flowers. Annuals are very easy to grow, flower profusely in a short time in numerous shades of colour. Due to variation in the season of growth and flowering, several species grow well in summer and rains, while the majority of annuals are cultivated in the winter months in the plains.

Sowing of Seeds

Seeds of annuals are sown in bed or shallow seed pan. The seed compost should consist of 1 part each of garden soil and finely screened dry leaf-mould. The compost should be slightly moist before sowing, otherwise the water will not soak readily and the seeds will float on water. Sow the seeds thinly and cover with a fine layer of screened leaf-mould. Protect the seed pan or the bed from rains after the seeds have been sown. When the seedlings start sprouting, exposure to sunlight should be gradually increased, so that before transplanting they become sufficiently strong to bear the full-day sun. The seedlings are severly damaged by fungul disease under conditions of high humidity and cloudy weather and it can be effectively controlled by the application of fungicide. Most of the seedlings become ready for transplanting at four-leaved stage. To see the beds of annuals in bloom in late December or early January, sow the seeds in September and transplant the seedlings in October. For a display of colour in early February, sowing should be done in early October. Double and large flowering Petunia, Cineraria, Carnation and Stock, which require about 5 months to come flower, should be sown earlier.

Preparation of Beds and Transplanting of Seedlings

The beds should be thoroughly dug at least 30-40 cm deep as the rain ceases and the soil is dry. The clods should be broken, soil pulverised and dug again to make the soil of uniform tilth throughout the depth of the cultivated layer. Well-rotted cow manure should be mixed at the time of preparation of the soil. It is gently pressed to keep the seedlings from the seed bed or pan, care should be taken to minimise the damages of roots. Seedlings will establish in the ground in about a week, if light watering is done when necessary.

Watering: After the seedlings have started new growth, they are watered profusely, weeds removed from these beds when it is just moist and the soil is allowed to dry before it is again flooded.

Manuring: Organic matter not only supplies nutrient to the soil but also greatly improves its physical condition. Cow manure is the best and easily available organic manure. Fresh or half-rotted manure should never be used just before planting of seedlings. Chemical fertilizers should be used in very small dose. Mixture of chemical fertilizers and organic manure in the form of solution, if applied in low concentration during the vegetative phase of the plants will greatly improve the growth and also the quality and quantity of flowers. Fertilizers can also be used on the soil as dry powder followed by flooding.

Types of Annuals: There are various types of annuals.

1. *Early Blooming Annuals:* Ageratum, Alyssum, Balsam, Browalia, Candytuft, Celosia, Dianthus, Gomphrena, Marigold, Salvia, Sunflower and Zinnia are some of the early blooming annuals.

2. *Shade Tolerant Annuals:* No annual can thrive in full shade, but where light is dappled and filters through a light tree the following will thrive. It must be remembered that the quality and quantity of bloom will necessarily be poorer.

   Ageratum, Alyssum, Begonia, Calceolaria, Calendula, Candytuft, Cineraria, Clarkia, Dahila, Delphinium, Godetia, Impatiens, Larkspur, Lobelia, Lupin, Myosotis, Nicotiana, Pansy, Phlox, Salvia, Verbena and Vinca are the shade tolerant annuals.

3. *Annuals for Rockery:* A rock garden in sun may not be entirely composed of cacti and succulents and patches of colour will brighten it. In a rock garden hardy types of dwarf annuals which thrive and flower in poor soil are usually selected. Ageratum, Alyssum, Arctotis, Brachycome, Nasturtium, Phlox, Portulaca, Verbena are some annuals suitable for this purpose.

Varieties

1. *Ageratum Conyzoides (A mexicanum) (Compositae) Ageratum:* Popular annual flowering plant for beds and pots. The height varies from 15-40 cm depending on the variety. It develops into well-shaped plant with dark green foliage and the dwarf varieties are compact and widely grown for edging and in a mixed border. Blue or its shades like azure blue, dark blue are normally selected for gardens; salmon, pink and white are not so popular. In the plains, seeds are sown from September to November and the plants flower in 2-3 months after sowing. Pinching of taller varieties is done to make the plants bushy. Removal of faded flowers promoted flowering for a longer period.
2. *Adthaea Rosea (Malvaceae) Hollyhock:* It is a popular flowering plant throughout the world and commonly known as hollyhock. In tropical climate, hollyhock is treated as an annual, while in the hills it is grown as a perennial. The height of the plant varies from 1.0 to 2.5 m. large flowers 8-12 cm across appear on long erect stem at the axils of the leaves. It is mainly used at the back of the mixed border or along a wall. Plants of shorter height also make good pot plants. The flowers open successively from below towards the tip of the spike and continues to bloom for a long time till the temperature rises high, when almost all other annuals dry up. A wide range of flower colour is recorded in hollyhock eg. White, cream, yellow, rose, lilac, purple, scarlet and crimson. The flowers may be single, double and semi-double. The plants grow better when directly sown in site and extra seedlings are removed to maintain a distance of 50-60 cm. Seeds should be sown in September-October to get bloom during January to April.
3. *Amaranthus Caudatus (Amaranthaceae) Love-Lies-Bleeding:* This species is commonly grown to pots or in beds as a flowering annual during winter

months and produces long pendulous plume like or headed spike, purplish crimson, light green and white in colour. The plants are quick growing and make a good display in groups in mixed border. The flower spike lasts for many days.

4. *Amaranthus hybrida (Princess Feather):* It is about 1 m in height and bears erect crimson red plume like spikes. Its cultivation is similar to that of *A.caudatus. A.tricolor splendens* is one of brightest foliage plant, about 50-90 cm in height. The leaves are deep scarlet marked with patches of yellow or bronze green. *A.tricolor rubra* has bright red leaves. The seeds are sown in February and March and seedlings grow into a well-shaped plant in about 2 months and continues to thrive a during the summer plant in about 2 months and continues to thrive during the summer and early rains. It grows well in pots and can be effectively displayed as foliage plants during the season.
5. *Antirrhinum Majus (Scrophulariaceae) Snapdragon:* Antirrhinum is one of the most popular annual grown in garden. The plants have dark green narrow leaves, bear tubular flower with spreading irregular lobes on long erect spikes at the end of main stem and branches. According to the height of the plant, antirrhinum is divided into 4 groups-tall (80-120 cm), intermediate (55-75 cm), dwarf (25-50 cm) and miniature (15-20 cm). There are varieties with ruffled or double flowers.

   Tall varieties with long spikes are very attractive at the background of a mixed border. The cut flowers are very showy and last long. Intermediate varieties in one or mixed colour are useful for planting in small beds or in patches in mixed border and as pot plants. Dwarf varieties have compact appearance and produces large number of smaller spikes at the tip of the branches. These varieties are commonly grown in beds, rock gardens, pots and window boxes. Seeds are sown during September to November. The soil should be porous but rich in organic matter and nutrients. In antirrhinum there is a wide range of colour except blue and with one or more colours of attractive shade.

   Tetra or tetraploid varieties with larger flowers and double flowered varieties are spectacular improvement in the recent years. Those plants have large funnel-shaped flowers, lobes widely open, slightly ruffled at the edges. F_1 hybrids or antirrhinum are vigorous plants and show better flowering than the common varieties.
6. *Arctotis Grandis (Compositae) African Daisy:* The species is commonly grown in the gardens of northern India. It grows best in porous soil, prolonged low temperature and less humidity. An appreciable improvement has been made in this species and the hybrids show wider range of colour and larger flowers. Normally *A.gradis* attains a height of 50-80 cms and bears white daisy like

flowers with lilac centre. Large flowered hybrids, developed as a result of crosses between the species and hybrids, show various attractive colours like organge. Arctotis is very useful for bedding and rock gardens and also as pot plants. The seeds are sown in September-October in the plains and in March-April in the hills. Flowers appear in about 4 months from sowing.

7. *Calendula Officinalis (Compositae) Pot Marigold:* Calendula is a popular annual in Indian gardens grown in beds and pots. The plant has a dwarf, rosette growth, bears flower heads 6-10 cm in diameter on stout stalk, 20-30 cm in length. Large double flowers are attractive, singles and semi-doubles bear large number of blooms, in various shades of yellow and orange. The plants grow well in rich porous but moist soil and flower in 2 1/2-3 months.
8. *Callistephus Chinensis (Compositae) China Aster:* Aster is one of the most popular flowering plants throughout the world. In the plains only annual types are grown, but many species of perennial asters are cultivated in temperature countries. They are grown extensively in beds for the attractive flowers of various shapes and colour. Cut asters last long and are used in vases and floral decoration. Height of the plants varies from 15 to 60 cm depending on the type and variety. There are various types *e.g.,* Giant of California, Ostrichplume, Commet, varying in the shape of the head and size, shape and arrangements of the florets. Wide range of colours include white, pink, rose, blue, scarlet, crimson, mauve, purple etc. Aster grows well in porous soil rich in organic matter in full sun. In the plains the seeds are usually sown in September-October, sowing in July-August is also possible in low rainfall area and plants bloom after 3 1/2 to 5 months depending on type, variety and climate.
9. *Celosia Plumosa (Amaranthacea) Cockscomb:* Cockscomb is a hardy annual and can be grown in winter, summer and rainy season.*C.plumosa* produces long silky feathery flower spike in various colours like red, orange scarlet, yellow and white. The height varies considerably from 80 cm in the tall ones to 25 cm in the dwarf varieties and intermediate ones having medium height in between.

   The feathery varieties with attractive shining spikes makes a good display in a mixed border. Dwarf types are good pot plants. The variety Cristata (C. plumosa var. cristata) produces orested flower heads, resembling the comb of cock in various colours like red, rose, organe, scarlet, bright scarlet, white, yellow, golden yellow etc. The height varies from 15 cm to 90 cm depending on the varieties.

   The seeds are sown in January-February, May-June and September-October for flowering in the summer, rains and winter respectively. Seedlings are transplanted at 4-6 leaves stage. Flowers appear in 2 1/2 to 3 months after transplanting.

10. *Centaurea Cyanus, Cornflower:* The plants are tall attaining a height up to 1 meter and are commonly grown at the back of mixed border. As the flowers last long, they are also used as cut flowers. Cornflower is a hardy winter annual in the plains and seeds are sown in September-October, flowers are produced in 3 to 31/2 months. There are varieties with pink, rose, maroon, purple, blue and white, mostly singly; double flowered varieties are also grown.
11. *Chrysanthemum (Compositae) Annual Chrysanthemum:* Annual chrysanthemums comprise of three species, *Chrysanthemum segetum* (Corn Marigold), *C.coronarium* (Crown Daisy) *and C. carinatum.* Corn marigold grows about 40 to 60 cm high and bears flowers 5.0 cm across in shades of white and yellow. Double flowered varieties are also available in this type.

    The crown daisy in much branched plants with finely cut foliage reaching a height up to 1 meter, size of the flowers varies from 2.5 to 4 cm and colour is usually in shade of yellow and white. Both single and double flowered varieties are found, but single is common.

    The tricoloured chrysanthemum (*C. carinatum*) are the best type in this group. The plants have finely cut leaves and bear large flowers variously coloured in white, yellow, scarlet, orange, purple, mahagon in the form of a ring. Disc appear at the centre and upper part of the petals. Annual chrysanthemums are free flowering and grow in all types of soil in sunny situation in beds or pots and used as cut flowers.
12. *Cleome Spinose (Capparidaceae) Spider Plant:* The plants are tall (1.0 to 1.3 m) much branched and spiny long protruding stamens and the four petals resemble spider, and the flowers are borne in large clusters at the end of the main stem and branches. In the plains it can be grown in winter and summer and flowers appear within 3 months after germination of seeds. Seeds are sown in September-October and January-February for winter and summer flowering respectively and seedlings are transplanted 40-60 cm apart. Colour of flowers are pink, rose and white.
13. *Cosmos Bipinnatus (Compositae) Cosmea, Cosmos:* It is a free flowering popular annual, used for bedding, in mixed border and as cut flower and grows well in sun and semishade in al types of soil. Though cosmos is known as a winter annual, it flowers early in the season, beginning from September and continues till March. The plants are tall (80-120 cm), much branched with feathery foliage and bear large flowers in pink, rose, crimson, lavender or white in colour. There are varieties with double flowers. Cosmos is best sown in situ though transplanting is not uncommon. Early sowing can be done in June-July in low rainfall areas and flowering commences in 2 1/2 months.

14. *Delphinium (Ranuculaceae):* Perennial delphinium though widely grown in the hills, is found to grow well as annual in the plains of northern India. Most of the cultivated varieties are hybrid between *D. elatum* and other species like *D. formosum, D. cardinale, D. belladona* and *D. nudicaule*. Perennial delphinium is tall (1.5 m) much branched with lobes and cut leaves, flowers appear on erect stout spike and colours include white, pink, lavender, lilac, purple pale and deep blue. Delphinium is ideal for mixed border, background and cut flowers. It is propagated from seeds, division and cuttings, the last two methods are only possible in the hill where it is grown as perennial.

    Larkspur, cultivated in gardens consists of two species *D. ajacis*, the hyacinth flowered and *D. consolida* the stock flowered larkspur. The plants are tall (80-120 cm), erect, branched with fine feathery foliage. Spikes of blue, purple, lilac, salmon rose or white flowers, singly or double arise in large number. Larkspur is excellent as cut flower and for growing in mixed border, along hedges and in semishade.

    Seeds are sown in October-November, in the plains germination is better in lower temperature. In the hills it can be sown in March-April or during August to October. Larkspur shows satisfactory growth and flowers in well-drained light soil and takes about 3 months to bloom.

15. *Dianthus Barbatus (Caryophyllaceae) Sweet Willam:* Sweet william has both annual and biennial strains. Annuals are grown in the plains, while biennials are grown in the hills. Annual varieties are 20-35 cm tall, have dark green foliage and flowers single or double in large rounded and showy clusters at the end of the branches. The flowers have a wide range of colours white, maroon, purple red pink, purple, rose pink, etc., self-coloured or zoned and variously coloured.

    Sweet william is widely grown in the plains and hills in beds, borders and are good for cut flowers. The seeds are sown in September to October and the seedlings will flowers during December to March. Biennial varieties take longer time to flower than the annual ones and are usually grown in hills.

16. *Dianthus Caryophyllus Carnation:* Carnation is one of the most popular annual grown in Garden; excellent for bedding, herbaceous borders, edging, pots and as cut flowers. In plains carnation is grown as a winter annual but cultivated as a perennial in hills. They grow and flower well in loamy soil and under full sun. Propagation by seeds and stem cuttings.

    Plants are medium tall, about 40-90 cm high. Stems grey green, narrow with swollen nodes. Leaves narrow, thick and often curves at the tip. Flowers single or double with smooth or fringed edged petals, sweet fragrance like cloves.

Terminal flowers are white, cream, yellow, pink, rose, red, scarlet, crimson, purple, maroon, orange etc. with selfs, striped, flaked and combination of different colours. Seeds are sown during August-September in the plains to get flowers from February, Flowering starts in 5 months from germination.

17. *Dianthus Chinensis, Indian Pink:* Dianthus is one of the most popular and beautiful garden annuals. The plants are 15 to 35 cm high and bear large rounded head of flowers, self-coloured in white, pink, rose, scarlet, crimson, violet, maroon, mauve, purple, etc. or zoned, spotted or blotched in various colours. There are town types of Dianthus-Heddewgii or Japanese pink and the fringed pink. The Heddewgii varieties are dwarf (15-25 cm) plants, free flowering, attractively coloured with single and double flowers having fringed and beautiful margin. The laciniatus varieties are taller (30-35cm), flowers single or double beautifully coloured with finely fringed petals.

    The pinks are excellent for beds, border, rock, gardens and cut flowers. They grow and flower well in loamy soil and full sun. seeds are sown during September-November and the seedlings bear flower about 3 months.

18. *Helianthus (Compositae) Sunflower:* The different varieties of sunflower grown for garden decoration can be grouped into two species-*Helianthus annus* and *H. dubilis.* Height of *H. annus* vary from 60 cm to 2.5 metres. Large single yellow flowered and large chrysanthemum flowered varieties are common. There are also large double flowered varieties plants of medium or dwarf height. Plants during the summer has chocolate, bronze or wine red flowers. A strain producing chocolate, bronze and wine red flowers are very attractive and popular in gardens.

    Miniature or Japanese sunflower belongs to the species *H. dubilis*. The plants are much branched, bushy, 120 cm in height and bear small yellow flowers. Sunflowers can be grown as summer and rainy season annuals for beds and borders. The seeds are sown from January-June and the seedlings transplanted after about 3 weeks. The seeds may be sown in situ.

19. *Impatiens Balsamina (Balsaminaceae) Balsam:* Balsam is a popular flowering plant in this country and widely grown in summer and rains. The plants are 20-60 cm tall usually branched and bushy and bear large number of single, semi-double or double flowers. The colours may be white, pink, rose purple, crimson, blue mauve and in various shades and also striped. Camellia flowered type is the best. Dwarf varieties (20-25 cm high) bear large number of flowers on the main stem and branches and make a beautiful display of colour.

    The plants grow well in rich, porous but moist soil and in high humidity in sunny places. They can also be grown in semishade. The seeds are sown in January-February for summer flowering and in May-June to obtain blooms during the rains. In case of Balsam, best method is to sow the seeds in situ and thin out the seedlings. Transplanting can also be done.

20. *Myosotis Alpestris (Boraginaceae) Forget-me-not:* The plants are 30-40 cm tall, much branched, bushy and bear cluster of small flowers at the end of the branches. Colour of the flower is white, pink, sky blue or dark blue. Myosot is flourishes well in pots and beds and can be grown in rock garden ad mixed borders. Seeds are sown in September-October in the plains and flowering starts after about 5 months.

21. *Petunia Hybrida (Solanaceae) Petunia:* Petunia is a popular and colourful annual commonly grown in beds, herbaceous border and in pots and window boxes. The plants may be 20-35 cm in height with trailing branches and produce many funnel-shaped flowers. There are various types of petunia, which show marked difference in growth, size, shape and colour of the flowers.

    i. Bedding—Hardy, free flowering type, bearing numerous small (4 cm across) funnel-flowers. The colours are white, pink, blue, violet, rose and red.

    ii. Grandifloras—This type is a comparatively low-growing plant and produces large (10-12 cm) single flower of attractive colour. The edge of the flowers is smooth, frilled or ruffled.

    iii. Doubles—Double petunias are magnificent flowers. The plants are widely grown in beds, border and pots. In the plains seeds are sown during August to October. The plants flower well in rich, porous soil and require 3-4 months to initiate flower. Doubles are also propagated vegetatively in the hills. In Japan, USA and Europe all double petunias are being produced by hybridisation.

    1. *Phlox Drummondii (Polemoniaceae) Pholox:* Phlox is a popular annual throughout the world. It is widely grown in beds, border, pot, boxes and rock garden. The types include *Grandiflora* with large flowers, *Nana compacta* having dwarf growth habit and star type producing star-shaped flowers. Many colours exist in grandiflora-white, light yellow, pink, rose, scarlet, crimson, dark blue, violet blue, violet etc. Nana compacta and star type produce compact bunch of flowers. Phlox is easy to cultivate and flowers well in rich soil. Seeds are sown in August-October and seedlings begin to flower with 3 to 3 1/2 months.

    2. *Pimpinella Monoica (Umbelliferae) Lady's Lace:* It is a tall growing annual (130-180 cm) with finely cut foliage and bearing small white flowers on loose heads at the end of the branches. Lady's face is grown in borders and in shrubbery. The flowers are commonly used for floral

decoration. Seeds are sown in September-October and the seedlings flower in about 3 months.

3. *Salvia (Labiatae) Sage:* Three species of Salvia are often grown as annual in this country-*Salvia splendens, S. coccinea and S. farinacea. S. splendens* is commonly known as scarlet sage. Height of the plant may vary from 20-70 cm depending on the variety. Bright scarlet flowers appear on long spike at the end of the branches, other colours include salmon, pink and violet.

   S. coccinea—Several varieties of this species bearing red, scarlet, pink or white flowers on terminal spikes are grown in Southern India.

   S. farinacea—The plants are 50-70 cm high, produces spikes of small blue or white flower. Salvia, particularly *S. splendens* is widely grown in garden for bedding, in border, pots and the dwarf variety in rock garden. It can be grown in semishade.

4. *Tagetes (Compositae) Marigold:* Marigold is a popular name among the flowers. There are two common types of marigold, African (Tagetes erecta) and French (Tagetes patula). It is widely grown in beds, pots, rock gardens and as cut flower for making garlands, bouquets and also for worship.

i. *African Marigold:* Many varieties of African marigold have been evolved, which show variation in the height of plant, shape and size of flower. The height of plants ranges from 60-100 cm or more, colours are lemon yellow, golden yellow and orange and recently another colour nearest to white is also available. Size of the flower may vary from 5 to 15 cm, carnation flowered type normally produces large flowers. Dwarf chrysanthemum flowered varieties are very useful for bedding. F_1 hybrids are superior to the common varieties.

ii. *French Marigold:* French marigold is a popular flower throughout the world, having a wider range of colour than the African type. It grows very easily, plant has dwarf, bushy growth (10-32 cm high) almost rounded in shape and bears large number of flowers of attractive colour. The flower colour may be yellow, golden, rusty red, mahagony, orange, deep-scarlet and often blotched, stripped or spotted in different colours. The heads may be single or double.

Although marigold can be grown throughout the year, mild winter is favourable for its growth and flowering. Hence, it is widely grown in the winter months in the plains and the seeds are sown in September-October. In the hills, sowing is done in March-April. French marigold flowers profusely in porous soil. While growth and flowering in African type are better in rich and moist soil.

1. *Zinnia ElEgans (Compositae) (Zinnia):* It is a popular flowering annual for summer and rains. There are many types and numerous varieties in zinnia varying in the plant height and size, shape and colour of flowers. The height may vary from 25-100 cm or more. Some popular types are Dahlia flowered. Burpeeana, Giant of California, Haageana. Thumbelina, Cupid, Lilliput and Joni Thumb. Colour of flowers may be white, yellow, pink, rose, orange, scarlet, violet, red, lilac, etc. and size of flowers may vary from 3-12 cm.

F_1 hybrids have vigorous growth and are tree flowering. The seeds of zinnia are sown from January to June for flowering during summer and rains. First flower bud appearing on small seedling is removed to induce branching. Zinnia grows well in light soil, rich in organic matter. The seedlings may be transplanted or the seeds sown in situ. The plants are highly susceptible to leaf curl virus in the plains.

Zinnia linearis is a dwarf and bushy plant, reaching a height of 20-25 cm. It grows well in winter and summer and produces numerous small white or orange flowers. The plants are very suitable for growing in beds, border and pots and continue to flower for several months.

Climbers and Creepers

Botanically, plants, which have special structure to climb on supports, are defined as climber. Climbers are very important ornamental plants and the beauty of any garden can be greatly increased by carefully selecting and planting them in a suitable place.

The growth habit and mass of colour give charming appearance. Climbers are very commonly used on arches and pergolas but in cities their utility enhanced for the purpose of screening the premises from adjacent houses and maintaining privacy. Bare walls can be most effectively decorated by growing colourful climbers on it. Fences and trellis also provide scope for the beautiful climbers to grow and display.

It is often felt that among the various types of ornamental plants least interest is taken on climbers and only few species or varieties are found growing in a locality. Even a good amateur gardener seldom knows more than a dozen climbers do and practical information regarding a large number of these plants are not readily available. Attempts have been made to include description of plants of horticultural interest and their utility and practical hits on cultivation of a large number of climbers suitable for tropical gardens. Another type of climbing plants grow spirally around another plant or support and do not possess tendrils, rootlets, shoots or thorn. These are known as twiners *e.g.*, Echites, Chonemorpha, Clerodendron. Creepers cannot grow vertically on their own *e.g.*, Morning Glory. All these types of plants are commonly known as climbers. Most of the flowering climbers require full sun for growth and flowering; few climbers, however, grow well in shade and flower *e.g.*, Thunbergia grandiflora, Clerodendron splendens, Jacquemontia violacea, Asparagus racemosus, Scindapsus, Philodendrons, Monstera deliciosa, Syngonium, Cissus discolor are grown for beautiful foliage in shade or partial shade.

Scindapsus (Pothos) are the most popular climber in shade and when grown on tree trunk, the leaves grow large with very attractive shades of green and yellow. Philodendrons are also grown on tree trunks in shady places. Ficus repens (Indian ivy) is the most popular and widely grown plant for covering bare walls or pedestrals. No support is needed as the roots from the node grip the walls.

Climbers which grow quickly and cover the space thickly are mostly used for this purpose. Railway creeper (Ipomoea palmata), Antigonon and Vernonia are commonly selected. Vermonia has neat growth but the flowers are not showy. Some heavy climbers with colourful bloom can be recommended. These are Clerodendron splendens, Derris scandens, Thumbergia, Adenocalymna calycina, Passiflora, Bignonia gracilis, Rhyncospermum, Bougainvillea, Ficus stipulata is also an ideal climber for screening shady place. Jacquemontia is a light flowering climber used for this purpose.

Ideal Climbers for Pergola: Quisqualis, Poiveria densiflorum, Banisteria laurifolia, Derris scandens, Clerodendron splendens, Petrea volubilis, Pyrostegia venusta, P. purpurea, Adenocalymna calycina, Bougainvillea.

Annual Climber: Important annual climbers are sweet pea and heavenly blue morning glory (Ipomea coerulea), Mina lobata with yellow and orange red spikes and Quamoclit vulgaris with white, pink and red flowers are also popular annual climbers. Thunbergia alata, (Black Eyed Susan) in white, pale and deep yellow with and without the dark eye are perennial but often grown as annual. Blue flowers are very common but white and mauve colours are also seen; the double blue flowering variety is the best.

Climbers for Pot Culture: For growing climbers in pots, light climbers or those with bushy growth should be selected. Climbers should be planted in large pots (20-30cm) and given suitable support to allow the plant to grow. Among the perennial climbers Clematis flammula, Bignonia purpurea, should be selected for pot cultivation. Bougainvilleas in pots are usually treated as shrubs.

Soil: Loamy soil with good drainage suits well to most of the climbers. Decayed cow or stable manure or leaf-mould should be incorporated in the soil of the pit about 50cm in depth and diameter in case of large and heavy climbers. For light climber, the pit may be 30cm deep and 25 cm in diameter. Bone meal and leaf-mould show very good result if cow-manure is not available. Distance of planting depends on the species or variety and the purpose for which the climbers are used. Quick growing heavy climbers are planted at a distance of 3m for screening, while light climbers for the same purpose may be grown at 1m. To cover fences or walls closer planting is recommended. Vernonia and ficus repens are planted at a distance of 15-30cm. As the climbers do not grow erect without support, main stem and the branches should be tied up with the support. This helps to grow and cover the space quickly. Like other perennial plants planting of climbers can be done any time of the year early monsoon is the best. Manuring encourages growth and flowering of climbers. Bone meal or hoof and horn once or twice in the monsoon will be beneficial to the plants.

Selection of Climbers: There are different situations in the garden like sunny, partial shade and different structures according to which suitable type can be selected from an available wide choice. The climbers for different situation are as follows:

- *For Sunny Situation:* Phyrostegia venusta, Quisqualis indica, Antigonon leptopus, Adenocalymma alliaceum, Bougainvillea sp., Campsis grandiflora etc.
- *For Partial Shade:* Clerodendron splendens, Petrea volubilis, Lonicera japonica, Trachalospermum jasminoides.
- *Heavy Climbers:* These climbers produce luxuriant vegetative growth and grow very vigorously. They cover large area and are suitable for covering bigger area *e.g.* Antigonon leptopus, Bignonia magnifica, Beaumontia grandiflora, Bougainvillea sp., Hiptage benghalensis, Quisqualis indica, Wisteria sinensis Pyrostegia venusta, Clerodendron splendens.
- *Light Climbers:* These climbers make sparse vegetative growth and remain light in spread. These climbers are suitable for limited spaces *e.g.* Lonicera japonica, Trachelospermum jasmenoides, Clitoria ternatea.
- *Climbers for Fragrant Flowers:* Jasminum grandiflorum, L. officinale, Trachelospermum jasminoides, hiptage benghalensis, Clematic penniculata.
- *Climbers for Pots:* Bougainvillea, Clitoria ternatea.
- *Making Hedge:* Clerodendron inerme and Bougaivillea.
- *Indoor Decoration:* Shade loving climbers like Pothos, Monstera deliciosa, Philodendron sp., Asparagus sp., etc. can be grown in pots and can be kept indoor also.
- *For Screening:* Vernonia elegnaefolia, Pyrostegia venusta.
- *Annual Climbers:* There are climbers, which are seasonal in nature and can be grown temporarily. *E.g.* sweet pea, trailing nasturtium. There are woody climbers, which can also be grown as annual like Ipomea lobata, Clitoria ternatea.

Clematis Flammula (Ranunculaceae)

A slender low-growing climber. Leaves usually bipinnate dark green, often trifoliate; leaflets narrowly lanceolate entire, glabrous 1.3 to 3.2 cm long. Flowers very fragrant, small 1.9 to 2.5 cm across pure white in loose panicle upto 20-30cm long, Sepals 4-linear oblong. Clematis flammula has less growth but flowers heavily on large panicles in the early rains. It is ideal for small gardens on low walls or wire nettings. It likes porous soil and partial shade. Propagation by layering.

Ipomoea Palmata, Railway Creeper: A perennial vigorous climber. Leaves 5.0 to 7.6cm in diameter, cut nearly to the base 5 to 7 lobes, entire, elliptic or lanceolate. Flowers purple, 5.0 to 6.3cm across often 3 flowered cyme. Corolla campanulate, colour deeper in the throat.

Ipomoea Purpurea, Common Morning Glory: It is also a quick-growing climber. Leaves entire, ovatecordate, 7.6 to 12.7cm long pubsecent on both surfaces. Flowers large funnel shaped, few on axillary peduncles, about 6.3 to 7.6cm across, white, pale blue or purple.

Passiflora Coerulea (Passifloraceae)

A vigorous climber. Stipule leafy, cordate, toothed. Leaves divided nearly to the petiole into 5 lanceolate or lance-elliptic segments 10.2 to 16.2cm across. Flowers solitary, large 7.6cm across, axillary. Bracts large ovate-cordate about 2.5cm in length. Petals 5 oblong-obtuse, pale pink in colour. Corona 5.0cm across, outer filaments blue at top, white in the middle and purple at base.

Passiflora Deulis: Woody, strong and spreading climber. Stem angular slightly downy or glabrous. Tendrils axillary, long simple. Leaves alternate large. 10.2 to 12.7cm long and wide deeply 3 lobed. Lobes ovate, glandular, toothed. Flowers solitary terminal or axillary, 3 leafy bracts, 5 lobed greenish with short crown. Petals 5 about 2cm long, oblong white often tinted with purple. Corona in several rows white in the upper half violet or purple below. Fruit globular, oblong, thickly pruple dotted.

Pyrostegia Purpurea: An evergreen climber. Leaves compound, leaflets 3 ovate to ovate-oblong about 5.0 to 7.6cm long mostly toothed bright green above paler beneath. Flowers 4 to 10 rosy purple in axillary cyme. Corolla funnel shaped about 5.0cm long, lobes 5 spreading about 6cm across. This ideal climber produces flowers in large bunches 6 to 8 times a year in the summer and rains and rains and should be grown on trellis or low walls. Propagated from layers.

Thunbergia Laurifolia: A shrubby climber. Stem terets, smooth except when young. Leaves opposite, ovate, oblong–lanceolate rounded at the base, entire or slightly toothed, 15cm long about 6.3cm wide. Flowers borne in axillary or terminal raceme. Corolla tube cylindrical, swelling above, cunnel-shaped with spreading large 5 lobed limb about 6.3cm across pale blue, white or yellow in the throat.

Shrubs: Shrubs are defined as woody, semiwoody or herbaceous perennial plants, branches arise from the base of the plants, and grow up to a height of about 0.5 to

4 metres. The plants are usually erect and bushy. Most of the shrubs are very hardy, require little attention and grow in all types of soil. Although some plants flower almost throughout the year, in most cases it is seasonal and selection can be made in such a way that some plants flower almost throughout the year, in most cases it is seasonal and selection can be made in such a way that some plants in the shrubbery will be in bloom in different seasons of the year. Shrubs flower regularly every year and produce flower of almost all the important colours-white, light yellow, golden-yellow, pink, scarlet, salmon, crimson, rose, violet, light blue, deep blue and some are sweet scented.

According to the requirement of sunlight for growth and flowering, shrubs can be classified into three groups:

- Show normal growth and flowering in full sun.
- Grow only in partial shade.
- Can grow well both in sun and semishade.

Most of the tropical flowering plants are sun loving and growth and flowering are greatly affected by shade and it is also observed that lack of morning sun shows much more adverse effect than the afternoon sun. Plants like Magnolia mutabilis, M.pumila, M.fuscata, species of Pentas, Olea fragrans. In full sun these plants show signs of scorching during the hot weather and do not thrive. Shrubs show great variation in height, tall shrubs usually reach a height of 3 to 4 meters and dwarf shrubs grow only up to 0.4 to 1 meter and thus plants can be chosen from a large selection for growing in various places according to their suitability and utility. In addition to their utility as flowering plants, shrubs are also used for hedges and screening. Acalyphas from the most colourful hedge, Dodonia viscosa, Duranta plumerii, Clerodendron inerme are usually grown as hedge plants. They are quick growing plants and stand severe and frequent pruning. Hibiscus rosa-sinensis, lxora chinensis, Bougainvillea, Barleria, Jasminum pubescens and Murraya exotica are also used as hedge plants and produce showy flowers. Cesturm, Bougainvillea, Jasminum pubescens, Murraya exotica, Jacquinia ruscifolia can be grown for screening. Hardy dwarf shrubs like Lantana depressa, L.sellowiana, Russelia, Euphorbia splendens, Gardenia fortunei, Vinca rosea, Turnera etc. are usually grown in rockery to add colour and also Adenium, Jatropha podagrica to break the monotony of succulents.

Selection, Planting and Cultivation of Shrubs

Selection of shrubs depends on availability and duration of sunlight at the site of planting, maximum height and spread of the plant, utility of plants *i.e.* foliage or flower, deciduous or evergreen and colour of flowers and season of flowering. It is always safe to plant shrubs during early monsoon. In the next two to three months of rains it will not only establish in the soil but will also start new growth and such plants thrive much better in the hot months of the year. Except hydrophytes garden plants are very susceptible to water logging and well-drained site should always be selected. Waterlogging for few days or very wet soil grown plant cannot be obtained in such a place.

In gardens, shrubs are usually planted in three ways:

a. In the form of shrubbery, on a long strip of land where taller shrubs are grown at the back, those with medium height in the middle and dwarf ones at the front. In this design upper portions of all the plants in the shrubbery bearing flowers become clearly visible. Colour of flowers and time of flowering are also considered so that the shrubbery remains colourful throughout the year with bloom and foliage.

b. In small garden or to break monotony of open space shrubs are often grown in-groups and in this case few plants of the same variety or of different species are selected.

c. Shrubs are also grown as a specimen plant or in only one row along the border of the garden or a wall.

In order to give a brighter appearance in the rockery dwarf and hardy shrubs are also planted. Distance of planting in a shrubbery depends on a size of the plants. Tall and bushy shrubs are generally planted at a distance of 1.5 to 2.5 ms. Shorter plants with less spreading habit can be grown at a distance of 0.5 to 1.5metre. Shrubs are watered in winter and more frequently in summer; slight wilting due to dry soil may cause defoliation and affect the growth of the plant. It is always advised to give thorough soaking to all plants then to sprinkle every day. Soaking of only surface soil encourages production of shallow root system, which not only prevents utilisation of nutrient from lower level of the soil but also makes the plant more susceptible to drought. Shrubs should be manured during the rainy season. Well decomposed farm yard manure can be dug in around the root region; use of bone meal or hoof and horn meal once or twice a year also shows excellent result.

Description and Cultivation

Acalypha: (Euphorbiaceae) Eng. – Copperleaf

Acalyhpa are colourful foliage shrubs used for many purposes. Most of the species are varieties have colourful shrubbery. These do best in full sun while in partial shade, growth is affected and colour becomes less attractive.

***Clearodendron Macrosiphon*:** It is very free flowering dwarf shrub 1.0 high and produced mass of white flowers in several flushes during summer and rains. C.macrosiphon should be grown in the front row of a shrubbery. Propagated by cutting or layering.

Duranta Plumeri (Verbenaceae)

Tall much branches shrub with axillary spines, about 2.5 to 3.0m high.

Var.: Alba - flowers white.

Var.: Variegata - flowers blue, foliage variegated.

It is usually grown as hedge and forms an impenetrable barrier.

Fortunella Japonica (Rutaceae)

Tall handsome ornamental plant when in fruit, reaches a height of 2.4m. Flowers small, scented. Flowers usually appear in the early rains and the fruits turn yellow in winter. It is very easy to grow and likes full sun. Propagation by air layering.

Hibiscus (Malvaceae)

These are the most useful flowering shrubs in Indian gardens. For species, H. mutabilis, H. schizopetalous, H. collinus and the varieties of H. rosa-sinensis are commonly grown in tropical gardens.

Pink Flowers

Ixora Chinensis Var: Rosea. A plant about 1.4m high. Flowers pale pink becoming reddish in age.

Ixora Rugosula : Compact growing habit with small crowded bunches of flowers, deep pink in colour.

Scarlet Flowers

Ixora Coccinea : A compact shrub about 1.5m tall. Flowers deep scarlet.

Ixora Chinensis : A small shrub about 1.5m high.

White Flowers

Ixora Barbata : A large shrub about 3m or more. Flowers white.

Ixora Undulata : A large evergreen shrub. Flowers white, sweet scented.

Lawsonia Inerme (Lythraceae)

Tall, often growing up to 3.6 to 4.6m.

Var: Alba petals light yellow

Var: Rubra petals rose on light green sepal.

Both the varieties grow tall and appear to be small trees with stout woody stem. Crown is much branched and spreading and bears large panicle of numerous small flowers in the summer and rains. Propagation by seed and cutting. Lawsonia is usually grown as hedge.

Changeable Rose-Hibiscus Mutabilis

Hibiscus mutabilis is a large perennial shrub with only a few straight branches. The green branches become woody with age and then the plant acquires the look of a small tree. In the early cold weather the plant produces single or double white flowers (10-12cms in diameter) that change to pink in the course of the day. Hibiscus mutabilis, commonly called the 'changeable rose' comes from South China. In the US the plant is known as the 'confederate rose' or the 'cotton rose'.

The plant prefers high humidity and a porous soil. It is host to some troublesome insects like scales and aphids. Manual removal of insects, though cumbersome, is a better method to maintain the health of the plant. Proper trimming after flowering is recommended for a second flush of flowering and to keep the plant in shape during the non-flowering season. The plant is propagated through seeds and cuttings. Seed raised plants have a tenacity to produce only single flowers. Being a large shrub with a good number of showy flowers, Hibiscus mutabilis is space filler and a colourful plant for the last row of a mixed shrubbery.

Cannas

Cannas are very important and popular perennial flowering plants and grow abundantly in the humid tropical regions throughout the world. They are very hardy and thus grown easily and successfully. The flowers have many shades of colour, appear throughout the year and make a wonderful display of colour which can hardly be surpassed by any other perennial plant. Cannas are frequently grown in private gardens but more extensively in the public gardens. As the flowers do not last long as cut flowers and are very common garden plants, visitors do not show much interest to pluck the flowers or pamage the plants. Cannas alone with their various shades of colour can make a garden quiet attractive. The genus Canna has about 50 species, native of Tropical America and Asia. The following three species of Canna are considered to be the parents of the present cultivated varieties.

1. *Canna Indica* (Indian shot): Rhizome stout, stem slender green 1 to 1.2 m high. Leaves oblong 45-60 cm long green, flower in simple raceme or in pairs; sepals-short, waxy; petals-pale pink, lanceolate; staminodes 3 red or rose about 5 cm long.
2. *Canna Flaccida:* Stem rather slender, 1.2 to 2 m high. Leaves ovate-lanceolate 20-30 cm long. Flowers few flowered raceme, pale yellow to sulphur; sepals-small; petals-linear-lanceolate, up to 7 cm long; staminodes-3 about 7 cm long.
3. *Canna Lutea*: Stem green 1 to 1.2 meters high. Leaves oblong or broadlanceolate, 25-40 cm long. Flowers on simple raceme pale yellow; sepals-oblong green white margined; petals-lanceolate pale yellowish white; staminodes pale yellow.

Classification: In India improvement of canna has chiefly been achieved through hybridisation. A large number of varieties have been developed as a result of hybridisation and selection and these varieties are now commonly grown in the gardens. Depending on the size and shape of the flowers, cannas have been classified as follows:

1. *Alipore Hybrids:* The selections are the results of 45 years of hybridisation and are a great improvement on the crozy types from which they are derived. The size of the flowers markedly increased with a wide range of colour.
2. *Bouquet:* In this class an ideal variety 'cupid' has flowers on closely branched spikes. The plants of this variety are dwarf.
3. *Candleabra:* This is a distinct break. The main flower stalk branches and as many as 8-12 spikes are produced instead or two or three.
4. *Crozy and Gladiolus Flower:* These are improved hybrids raised by Anne in about 1850 and Vilmorin in 1880.
5. *Dreadnaught:* A great advance on the ordinary crozy or gladiolous flowered canna both as regards the individual flowers and bunches.
6. *Dwarf:* This type include varieties which does not exceed the height of 70-80 cm at any time of the year and are very effective for breeding purpose.
7. *Giant or Orchid Flower:* Originated in Italy and was very popular for many years. The large blooms of silky appearance resemble the Flag Iris, but not very hardy.
8. *Miniature:* This is small flowering type derived by crossing a society's dwarf hybrid with Canna indica. The spikes are neat and compact.

Cannas can also be divided into several classes according to the colour of the flowers.

1. *Selfs:* without spots or margin, one colour only.
2. *Spotted*: usually a shade of red on cream or yellow ground or red spots on orange of red ground.
3. Striped red on a cream or a yellow ground.
4. Margin yellow.
5. Margined with a darker shade than the ground colour.
6. Flaked red or orange on a paler ground.
7. Splashed orange on a deeper ground.

The shades of colour cover a wide range from creamy whit through yellow, orange and pink to an intense maroon red. There is no blue or purple colour in canna. There are also few varieties with dark leaves usually reddish purple, also producing dark red flowers. In one variety, however, the leaves are striped yellow.

Propagation: Canna is commonly propagated by rhizome. Seeds are used for raising new varieties through hybridisation. The seeds have hard seed coat and take a long time to germinate, if the seeds are not treated with hot water or rubbed on a sandpaper.

New varieties result by hybridization. For this purpose they are raised from seeds. Bud sports also produce new varieties. The seeds are of a small pea size and black in colour with a hard seed coat. The seeds are scarified by filing, or by rubbing with a sandpaper or a small portion cut off with a sharp razor blade without injuring the embryo which is inside the seed. This treatment ensures the better germination of seeds. Sometimes, soaking of seeds into water overnight or keeping them in cowdung before sowing is also useful obtaining better germination.

Cultivation: The rhizomes are collected in the month of May and then planted to get the full benefit of the Monsoon and a strong clump forms before the winter. The soil is prepared by digging at least 50 cm deep, breaking the clods and mixing in fresh stable manure at the rate of 100 kg to a bed of 10 sq. metre. The rhizomes are buried 3 cm below the surface of the soil and thoroughly flooded. If the weather is hot and dry, shade is provided for a few days. Planting is done at a distance of 30-40 cm between the rows and the plant. Within six weeks after planting first flower spikes appear which should be removed to. Encourage better growth. The bed should be kept free from weeds by frequent weeding. After the rain ceases and the soil thoroughly dries, it is loosened and allowed to dry. Digging and watering of the bed are necessary at least once a month in the winter and more frequent flooding is recommended to keep the soil moist. After the first flush of flowering is finished in December, leaf-mould or cowdung manure should be applied to the beds to ensure good flowering in the second flush from April to June.

Canna is replanted every year. The rhizome is lifted with a forked hoe, cleaned properly and the top growth cut leaving a stem of about 15 cm. The clump is then split taking only the strongest: root and stored in a shady place for a week or so before replanting. The canna beds should be kept clean by collecting faded flowers and removing dry leaves.

Cannas can also be grown in 30-35 cm pot and the compost should consist of 2 parts stable manure and 1 part garden soil. The potted plants should be replanted every 6-9 months as the roots become pot bound.

Pests and Diseases

Cannas are hardy plants and are particularly free of diseases. Caterpillars occasionally attack the leaves specially during the rainy seasons and bettles sometime destroy the flowers. Application of insecticides like Malathion, Basudin, etc. will keep away the insects.

Varieties of Cannas and Their Flower Colour

Name of the Variety	Flower Colour
American Beauty	Handsome orange scarlet flowers.
Apricot	Buff-yellow base overspread with salmon pink.
Aurora Borealis	Canary-yellow with rose pink centre, beautifully rayed.
Carmine king	Bright carmine-red with yellow centre
Cleopatra	Orange terracotta, Light purple leaves
Dorris	Pale salmon-pink Flowers very pretty.
Golden Wedding	Yellow
Louis Cayeux	Large flowers with bright rosy scarlet colour.
Mrs Herbert Hoover	Beautiful flowers with deep watermelon pink colour
Mrs Pierre S.du Pont	Very charming flowers with crinkled edges and light watermelon pink colouration.
Rosamond Coles	Dark reddish-orange with deep orange yellow border.
Rosea gigantea	Very large flowers with soft rose to carmine-pink colour.
Statue of Liberty	Plants with bronze coloured leaves and blazing flame-red flowers.
Susquehana	Dwarf plants with rose-pink flowers.
The President	Beautiful large flowers with rich growing scarlet colour.
Yellow King Humbert	Bi-coloured flowers. Bright yellow petals marked with crimson dots.

Bulbs and Tubers

In horticulture, the word bulb includes underground modified stems which are used for propagation *e.g.* bulb, corm, tuber and rhizome. Plants with tuberous roots are also grouped as bulbous plants. A large number of these plants producing attractive flowers are grown in the hills and are commercially important plants in floriculture. Many types, flower well both in plains and hills but the season of growth and flowering may vary.

A bulbous plant has normally three phases during a year—growth, flowering and dormancy. Vegetative growth may precede flowering as in Gladiolus or flowering starts before the leaf emerges, *e.g.* Amaryllis, Haemanthus. After growth and flowering, the plants in most cases, enter into rest period and the duration of dormancy varies with the type of plants and environmental conditions like temperature and humidity.

Cultivation

Bulbs prefer loam or sandy loam soil. In stiff clay, rooting is delayed and too much moisture often causes rotting. If the soil is not perfectly well-drained, the bulbs may be planted on a bed of sand. Watering is not required after planting as the fleshy underground stems contain sufficient food materials to develop initial growth of root and shoot. Before root formation, watering prove injurious to the bulbs and helps in rotting.

Bulbs, corms, tubers and rhizomes are used for vegetative propagation. Various methods and structures used for propagation are as follows.

- Division of clump—Rhizomatous plants the Canna, Alpinia can be divided into small clumps.
- Offsets produced laterally are separated for multiplication.
- Cormlets or bulblets produce new plants.
- Scales of Lilium are separated and used for propagation.
- Pieces of bulb disc with leaves or rhizome also develop new plan
- Bulbil arising from the axils of leaves can produce new plants.

Bulbous plants are grown in beds, shrubbery, herbaceous border, grassy land and in pots and bowl. Bulbs, corms, rhizome or tuber are usually planted when they have shown signs of sprouting after dormancy.

The planting materials are normally placed deep in the soil and the soil around it is gently pressed. The depth of planting varies with the type or plants and size of bulbs. For most of the bulbous plants grown in tropical garden, the planting depth is between 3-10 cm.

Most of these plants sunny situation, while Eucharis, Zephyranthes thrive better in semishade; few types, *e.g.* Pancratium, Haemanthus, Zephyranthes grow and flower well in sun and semishade. Bulbs, corms and tubers should not come in direct contact with fresh organic manure, rhizomes are not usually affected.

Some Common Bulbous Plants

1. *Amaryllis (Amaryllidaceae):* Amaryllis belladona is the only species of the genus, a native of South Africa. It is a plant with strap-shaped, glossy, green leaves and producing few large funnel-shaped flowers on a stout stalk, from March-May. Leaves appear at the end of the flowering season and die out before the winter sets in. In warm humid region the plants retain leaves throughout the year and bear flower during the season. Offsets, developed from the original bulb are separated for multiplication. Bulbs are planted 5-8 cm deep in December-January and the plants grow well in rich and porous soil. Amaryllis is not disturbed for several years at one place and the plants continue to produce new bulbs and flowers. It is very suitable for planting in border, shrubbery and in pot. Although rose or rosy pink colour is common, white to purplish colour are also found.
2. *Caladium (Araceae):* The genus comprises of about 16 herbaceous perennial species with tuberous rhizome. They are a native of tropical America and are widely grown in warm humid climate for the beautiful and attractive foliage produced during the rainy season. Some of the important species are Caladium

bicolor, C. humboldtii, C. picturatum, and C. schomburgkii. Numerous varieties have been raised particularly form C. bicolor.

Leaves are usually peltate-segittate, stalks variegated, blades are with very many shades of colour, *e.g.* variegated green, blue green, dark green, light green, spotted with white, red, transparent white, etc. leaf veins may be red, silvery or green. Leaf margins are coloured with purple, white, yellow or red. In some species, shape of the blade is lanceolate-sagittate. Flowers are unisexual.

The pot compost for planting Caladium should be prepared by mixing loamy soil with leafmould and well-decomposed cow dug manure. It prefers moist but porous soil. Application of liquid manure at frequent intervals enhances growth and improves colour of foliage. Caladium is propagated through division of the tuberous rhizome and rarely by seeds. It is planted in summer and leaves show a fine display of colour in the rainy season.

The leaves start fading in autumn and water is gradually withheld until they have withered. In dormant condition, the rhizome may be allowed to remain in the soil or dug out and stored in a cool and dry place.

3. *Cooperia (Amaryllidaceae):* The genus is named after Joseph Cooper, an English gardener, comprises about 6 species native of North America, and differing from Zephyranthes by the long perianthtube and erect anthers. They are tender bulbous plants with the habit of Zephyranthes but blooms only at night. Important species are C. drummondii and C. pedunculata. Flowers are fragrant, white, and sometimes tined with red or pink.

   Flowers are solitary, the perianth subtended by a bract like spathe. Leaves are long, narrow flat and twisted and appear along with the flower. It is cultivated in a semishady location and in sandy loam soil. Addition of well-rotted compost, sand and charcoal dust in beds or pots is beneficial. Cooperia is propagated by bulbs. It is planted during spring in rock garden, border or in pots and starts blooming during May-June.

4. *Cooperanthes (Amaryllidaceae):* This genus is a product of intergeneric hybrid between species of Cooperia and Zephyranthes, first raised in 1900 by Percy Lancaster at the Agri-Horicultural Society, Alipore, Calcutta. Hybrid between Cooperia oberwettii x Zephyranthes robusta is known as Alipore Beauty, is probably the best known Cooperanthes. Flowers are light lilac, rose or white.

It grows well in well-drained sandy loam soil, rich in organic matter and in semishade. Cooperanthes are propagated by bulbs. They are planted in the spring and the plants start flowering by the end of summer. They are grown both in bed or in pot and are good as cut flowers.

1. *Crinum (Amaryllidaceae):* The genus Crinum, comprises more than 100 species of large and showy flowering bulbous plants. They are closely allied to

Amaryllis and distinguished by the longer perianth tube. The species cross freely and many fine hybrids of Crinum, between Crinum and Amaryllis and Crinum and Hymenocallis are known. Some of he important cultivated species and hybrids are C. longifolium, C. mooriei, C. powellii, C. variabile, C asiaticum, C. augustum, C. careyanum, C. yuccaeflorum, C. giganteum, C. zeylanicum. Flowers are usually white or in shades of red and purple.

The stems arise from the tunicated bulbs with a more or less elongated neck. Leaves are large, about 150 cm long and 12-15 cm wide, evergreen or deciduous, depending on the species. Flowers are regular, often highly scented, tube narrow, with six segments and usually funnel-shaped. In some species flowers are 30 cm long and 15 cm wide.

Most of the species prefer shade or semishady location for planting. Crinums usually have large bulbs, sometimes as much as 60 to 80 cm long with numerous fleshy roots. If planted in beds, the soil should be dug to a depth of two to three feet, mixed up well with sufficient quantity of rotten cow dung and compost. It can also be grown in large pots containing soil rich in organic matter.

Bulbs should be of good size and planting is done to a depth below the ground level, twice the size of the bulbs. The plant will grow and flower for years, if watered in the summer months and top dressed with fresh loamy soil around new vegetative growth. Bulbs are planted in April to get flower during the rainy season.

2. *Eucharis Grandiflora (E.amazonica) (Amaryllidaceae):* Eucharis is an important bulbous plant, popularly known as 'Amazon Lily'. It prefers semi shade and flowers better in plains than on the hills. The bulbs are globular in shape, leaves large lanceolate. The flowers are about 7 cm across, white, sweet, scented, 5-7 blooms appear on a stalk. Eucharis grows better in pots than in the ground and the compost should be rather rich and heavy instead of sandy, but must not be often disturbed. It flowers in the summer and rains and the leaves begin to wither before the winter.

3. *Gladiolus (Iridiaceae):* Gladioli are among the most beautiful flowers, blossoming from October to March in plains and during June to September in the Hills. This genus comprises more than 150 species of perennial herbs with base of stem swollen into a corm. Most of them have their origin in South Africa, although a few originated from Europe. The species in cultivation and of considerable importance include: G. cardinalis, G. childsii, G. colvillei, G. gandavensis, G. lemoinei, G. nanceianus, G. nanus, G. primulinus, G. psittacinus, G. purpureoauratus, G. saundersii. The numerous crosses have evolved large number of varieties of varying characters and it is almost

impossible to keep up with and enumerate the different varieties of gladioli. Wide range of attractive colour shades may be grouped into white, yellow, cream shaded pink, orange, pale pink on white, pink on cream apricot, salmon, cherry red, orange-scarlet, crimson, purple, mauve and violet, etc.

Gladioli prefer sunny situation with light sandy soil, with pH between 6 to 7. If the soil is heavy, addition of river sand and charcoal improves the soil condition. They grow well both in pot and in beds and the magnificent spikes brighten the garden and room as cut flower. Two parts sandy loam soil mixed with one part of each well-rotted cow dung manure and leaf-mould and a handful of bone meal is recommended for pot culture. If planted in beds the ground should be dug deep and well-decayed compost at the rate of 4 kg per meter mixed thoroughly and left for sun drying at least for a fortnight. At planting time fresh manure should not be used, but a dressing of bonemeal, superphosphate and wood ash may be given with advantage. In the plants, planting is usually done during September-November and in the hills from April-June, but the flowering season may be extended by early and late planting.

Gladioli are propagated by seeds, corms and cormlets. New varieties are raised from seeds. Seeds germinate freely and the seedlings grown carefully, will flower in the second year. Large flower spikes, however, develop after 3-4 years when the corms attain a good size. Gladioli corms are planted 8 to 10 cm deep into the soil, at a distance of 20-30 cm between the rows and the plants. Application of liquid manure, once at the vegetative stage and again after the formation of flower buds has been found very effective.

The flowering spikes appear in 60-90 days after planting, the flowers continue to open in succession from below upwards and the open flowers remain fresh for a number of days. After flowering, the leaves begin to turn yellow and wither. The plants are then lifted with the corms and cormlets and kept in a dry shady place for a week for drying. The corms are thoroughly cleaned, the cormlets separated and stored on a layer of sand in a dry, cool, airy and shady place until the next planting season. The corms should be examined regularly and those showing sign of rotting or fungus growth should be removed. They are also stored in cold storage but in a dry atmosphere.

4. *Gloriosa (Liliaceae):* As the name implies, the genus Gloriose means 'full of glory', popularly known as climbing or creeping lily. They are tall creeping plants, support themselves by means of tendril which arises from leaves. This genus comprises about six species of rhizomatous plants native of Africa and tropical Asia. Important species, commonly grown in the gardens in India are G. superba and G. rothschildiana. Leaves are oblong, lanceolate. Flowers are

showy on long pedicels in leaf axils, perianth of 6 distinct long segments; stamens six with versatile anthers.

Well-drained soil in a sunny location is ideal for planting Gloriosa and the attractive flowers make a fine display of colour when trained on bamboo frame work or low trellis. Gloriosa is vegetatively propagated from rhizome, which may be cut in pieces and planted to a depth of 3-4 cm in April-May.

The plant flowers in July-August. After flowering is over the plant begins to wither and rhizome becomes dormant. If left undisturbed. Gloriosa continues to produce flowers for several years in the same place. Application of liquid manure once at the active phase of vegetative growth and another just before flowering is recommended.

5. *Haemanthus (Amaryllidaceae):* Haemanthus is one of the popular and attractive bulbous plants, commonly known as Football Lily or Bloody Lily. The genus comprises nearly 60 species, native of South and tropical Africa. The flowers are red, crimson, scarlet, pink and some species are white or pale green in colour. Some of the important species are H. multiflorus, H. magnifica, H. lindenii, H. albo-maculaius, H. coccineus, H. tigrinus, and H. candidus.

   Bulbs are usually large with thick skin. Stems green, short, thick and fleshy. Leaves are usually large and luxuriant, turn yellow and dry in the winter months, the scape is sometimes curiously coloured. Inflorescence is a dense, many flowered umbel, perianth straight and erect with a short cylindrical tube. Flowers are showy and produced in ball-like heads. Fruits berry like, indehiscent. Flowers often appear before the leaves or sometimes simultaneously.

   To plant in beds, the soil is deep dug and mixed with well-rotten compost and planted 40 cm apart. It grows well in pot and small pot is preferred. Application of organic manure in the rains and liquid manure before flowering is beneficial and the plants will continue to flower for several years in the same place.

   They are propagated by offsets, which should be detached from the mother plant during spring. The bulbs are planted in pots or in beds during spring season to get bloom in summer and early part of rainy season.

6. *Hedychium (Zingiberaceae):* Hedychium consists of 40 species of rhizomatous herbs native of Asia and several species are grown for their beautiful and fragrant flowers in both plains and hills. It prefers semi-shade and moist soil and produces many flowered large spikes during July-October. The leaves die before winter.

7. *Hemerocallis (Liliaceae):* Hemerocallis is popularly known as Day Lily as the flowers last for a day and the blossoms fade at night. This genus consists of

more than a dozen species and they are mostly native of China and Japan. The plants are very hardy, stout rooted, glabrous, perennial herbaceous plants, admired for their showy blooms.

The leaves are almost grass like, 2 ranked at the base of the scape. Flowers are lily like, large funnel shaped, yellow or reddish orange or brown in colour. Though the individual flower last for a day, but many flowers open successively to keep the lasting beauty for a long period.

Hemerocallis grows in wide variety of soil. While planting, the ground should be dug to a depth of 50 to 60 cm and mixed up with well-rotten compost. They take a year or two to establish properly and should not be disturbed very frequently. Planting is done in February to March and flowering continues from May to August.

Propagation is by division of the clumps and also by seeds. Some of the species are self-sterile and seeds can be obtained easily by crossing two species. Many new-varieties with attractive large flowers of various colours have developed by hybridisation. After 4-5 years, Hemerocallis clumps may be forked out during January-February separated and can be replanted again in the planting season.

8. *Hippeastrum (Amaryllidaceae):* Hippeastrum, commonly known as Royal Dutch Amaryllis is one of the finest flowering bulbous plants. The flowers of Royal Dutch Amaryllis are larger, open widely, about 15-20 cm across, tube is shorter and the plant develops larger bulb than Amaryllis belladona. The spectacular flowers have various shade of bright colour and wide range of variation exists in varieties developed by hybridisation. Hippeastrum produced by Ludwig are famous throughout the world. A double flowered variety is also available.

   The leaves are broader than Amaryllis and do not dry out in the plains. The bulbs are lifted in November-December, stored in dry cool place for a few weeks and replanted in January-February. Hippeastrums develop flower during March-April and application of liquid organic manure before emergence of flower stalk improves the size and quality of flowers.

9. *Hymenocallis (Amaryllidaceae):* This genus comprises about 40 species of bulbous plants native of South America (one in Africa *i.e.* H. speciosa) cultivated for beautiful fragrant flowers, consisting of narrow green white petals with large coronas and protruding stamens. Some of the important species are H. macrostephana, H. harrisiana, H. speciosa, H. caribaea, H. calathina, H. rotata, H. amancaes, H. tubiflora. Flowers are white, except H. amancaes which is bright yellow.

   Well-drained soil in sunny location is good for planting Hymenocallis. The bulbs are planted just below the surface of the soil during February to get the bloom in the month of May. The soil should be prepared by mixing 2 parts of loamy soil, one part of each well-rotten cow dung manure and leaf-mould. Watering should be given freely during the growing season. So long the plants remain healthy, transplanting or shifting from pot to pot is not recommended. Whenever necessary bulbs are dug out during September -October and stored in dry place for planting in February next year. Hymenocallis is propagated by bulbs.

10. *Nelumbo (Nymphaeaceae):* Nelumbo or Nelumbium, popularly known as Lotus is a flower of national importance in India. The genus consists of two species– the one bearing yellow flower and the other produces white or pinkish flower.

    The plant has rhizomatous stem, large peltate leaves standing well above the ground and it spreads rapidly in shallow water or wet soil. Like Nymphaea, Lotus prefers soil rich in organic matter and flowers profusely in summer and rains. Lotus do not grow well in lily pools and so tank is ideal for it. N. lutea bears scented yellow flowers 15-20 cm across N. nucifera is common in India. The flowers are white or pink, 20-25 cm in diameter. Lotus is propagated by rhizome or seed.

11. *Nymphaea (Nymphaeaceae):* The Water Lily is the most popular flowering plant in water garden. The genus comprises of 40 species of aquatic, rhizomatous plants widely grown in tank, lily pool, streams, lakes and in suitable earthenware containers.

    The leaves are oval or round in shape, floating on the water or standing above the water surface. The flowers are very attractive, show various colours: white, yellow, pink, red, blue and shades of colours are found in the varieties developed by hybridisation.

    Water Lily prefers clay soil rich in organic matter and sunny location. The rhizomes are planted in soil in the bed of the tank or lily pool or in baskets or pots filled with rich clay soil and then placed in the tank in March-April. In the gardens in cities and towns water lily is often planted in large earthenware container filled with water and a layer of soil. In several species and varieties do not develop in the winter months and the rhizomes remain dormant. Vegetative growth starts in spring and flowers appear during the summer, rains and early winter.

    In some species flowers open in day time while in other blooms open at night and close during the day. The species of Nymphaea commonly grown in water garden are as flowers: N. caerulea produces light blue flowers, 7-10 cm across, day blooming. N. lotus, the night blooming water lily, bearing white flower 12-15 cm across. N. odorata, a white flowering and fragrant water lily. The flowers are 8-15 cm across. N. pubescens, has white flowers, 8-12 cm across.

N. stellata, a species commonly found in the warm humid region of India. It produces pale blue flowers about 10-16 cm in diameter. Nymphaea is propagated from rhizomes, bulbils and seeds. Most of the species and varieties thrive and flower well at water depth between 50 to 150 cm.

12. *Polyanthes Tuberosa (Amaryllidaceae) Tuberose:* Tuberose (Polianthes tuberosa), a native of Mexico, is widely grown in the plains of India and blooms profusely during the summer and rains, flaunting its fragrance outdoors and indoors. Most artistic garland, floral ornaments, bouquets and buttonholes are made from these flowers. The flowers remain fresh for days together and bathe the atmosphere with their sweet pleasant fragrance.

    The tuber is bulb-shaped and the plant is commonly classed among the 'Bulbs'. The leaves are 70-80 cm long, narrow, linear and radical, bright green in colour. The flowering stalk which emerges from the centre of the cluster of leaves is about 80-120 cm long bearing successively smaller long pointed clasping leaves, uppermost ones are much reduced and bract-like. The flower buds are tubular. Flowers are 5-6 cm long, borne in pairs in an open spike pure wavy white, highly fragrant, tube 2.5 to 3.0 cm long, slightly bent near the base, expanding widely where it meets the oblong obtuse segments.

    There are three types of tuberose in cultivation-'single' with one row of corolla segments; 'semi-double' bearing flowers with two to three rows of segments and 'double' having more than three rows of corolla segments. Though there is no popular named variety in tuberose. 'The Pearl' is known to be a variety in the double flowered type. A variety with variegated leaf bearing single flowers is also grown. Single flowered type is more widely cultivated than the other types.

    The bulbs remain dormant during the winter months in places where the temperature is low and if early planting is desired, the dormancy can be successfully broken by dipping the bulbs in 4% thiourea solution for one hour. Normally tuberose begins to flower in 90 –95 days after planting. It flowers during the summer and rains (April-September) in the plains of eastern part of the country and from May to July on the hills, while in milder climate tuberose flowers well throughout the year.

    Bulbs having diameter 2.0 and 2.5 cm show satisfactory growth and flowering. The average life of a flower spike is about 10-15 days in situ, while that of an individual flower varies from 4-6 days. Vase life of a spike varies from 7-10 days, depending on the environment and change of water. The spikes remain fresh for a longer period, if kept in 4% sugar solution.

    Tuberose can be successfully grown in pots, beds, borders, shrubberies and rockeries. Propagation is by means of seeds and bulbs. Vegetative propagation is commonly practiced and desirable too, because such plants produce better flowers within a short period after planting.

The land should be thoroughly cultivated until the soil comes to good tilth. A good amount of rotted cow dung or farm yard manure should be incorporated with the soil at least 10-15 days before planting. The bulbs are planted 4-5 cm deep in beds in March-April in the plains and in May-June on the hills. The planting distance between the bulbs is 10-15 cm, while the rows are spaced 25 cm apart. For pot cultivation, 1-2 bulbs are planted in a 20 cm pot.

A fertilizer mixture containing 6 gms of urea, 16 gms each of superphosphate (single) and muriate of potash per square meter has been found to show satisfactory growth and flowering. The above mixture should be applied in two equal doses – the first dose before planting and the second one, 4 weeks after sprouting of the bulb. Commercial growers dig out bulbs 3 years after planting. They are stored in a cool, dry and shady place and planted again in the following spring.

13. *Zephyranthes (Amaryllidaceae):* Zephyranthes has a fanciful meaning *i.e.* the 'West-wind Flower' commonly known as Zephyr Flower or Fairy Lily. The Genus has about 50 species of bulbous flowering plants, native of warmer region of America. It is related to Habranthus, Pyrolirion and Hippeastrum. Some of the popular species are Z. candida, Z. citrina, Z. grandiflora, Z. rosea, Z. macrosiphon, Z. tubispatha and Z. verecunda. The flowers are with many colour ranges *i.e.*, white, yellow and various shades of pink.

Zephyranthes is a hardy bulbous herb. Leaves are filiform or linear or may be strap shaped. Flowers solitary, peduncle elongated slender, hollow; perianth funnel-shaped, erect or slightly inclined, stamen 6, stigma 3-fid. It is commonly propagated by separating the bulbs, sometimes through seeds. Bulbs are planted in spring, 30 to 40 mm deep and 2 to 15 cm apart. They flower during summer and rains and make a very colourful display especially in rains. Top dressing of organic manure once a year in a rainy season will promote flowering and the bulbs may be left undisturbed for many years.

Hedges

Hedge is an important adjunct to every garden. It is necessary to demarcate the garden from public road and from adjacent gardens particularly in flat gardens. Besides its utility as screens, hedge is necessary between flower and vegetable garden, to provide a background for annuals if shrubbery does not exist and also to enclose a rose garden or a children's corner in a park or public garden. A windscreen is also wanted when strong breezes, hot or cold blow from one direction during certain seasons of the year and frosts. It is useful to screen the manure pit, potting area, servants' quarters and other unsightly views.

Growing the Hedges: The majority of hedges are propagated by cuttings but where seed can be obtained the cost of raising a large number of plants in their delicate stage is difficult. If this cannot be carried out sow at the following rates per 100 metres; *Duranta* 1 kg, Inga dulcis 1kg, Jaint (*Sesbania*) 150 gms, Mendi (*Lawsonia*) 250 gms, Dodonea (*Sanatta*) 250 gms. Hard seed such as that of *Acacia modesta* should be soaked in warm water for 24 hours.

Selection of Plants

Shrubs: A hedge of *Breynia rhamnioides* is often seen in gardens and is very ornamental when in berry. *Bauhinia acuminata* in spite of its large leaves makes an effective display of white flowers even when brought to a shape. *Caesalpinia pulcherrima* ages fast when severely pruned but if trimmed slightly it provides tall hedge. *Clerodendron inerme* forms neat hedge or edge of varying height for m20 cm to 2 metres. It forms a compact hedge between 20-120 cm. C. inerme is not touched by cattle and white ants leave it alone.

Duranta in its many varieties holds first place in Bengal, *D. variegata* being very ornamental. Next comes *Lawsonia* (Mendhi) which is better in dry localities than in lower Bengal. *Dodonea viscosa* (Sanatta) is also an up country hedge but this latter should not be cut back as it is liable to die off in three or four years. *Nerium* and *Tecoma* are two others that might be used. *Dracaena fragrans* and *D.sticta* are good hedges for a shady spot. *Hibiscus* of several kinds form excellent hedges but *H. liliflorus* and *H. schizopetalus* are chiefly used.

Ehretia buxifolia as well as *Serissa foetida* and *Malpighia coccigera* are small leaved evergreens for shady situations. *Polyscias* (Aralia) of many kinds. *Codiaeum* (Croton) *Eranthemum* and *Graptophyllum* are ornamental shade loving hedge plants. Crotons of the narrow leaved and small foliage varieties are better hedges than those with large foliage. *Plumbago capensis* and *P. zeylancia, Meyenia* (Thumbergia) erecta and *M. affinis* are dwarf types of flowering shrubs that can be used for a hedge. In Bangalore we have seen *Plumbago capensis* cut down to 15 cm and in bloom. Other shrubs that can be used for hedges are *Caryopteris, Mussaenda, Rhapiolepis* and *Rondeletia.*

Euphorbia pulcherrima, if backed or having a supporting hedge in front can provide a fine mass of colour in the cold months. The last pruning should be done in August –September. *Jacquinia ruscifolia* is an evergreen hedge with orange red flowers and needled–tipped leaves and ideal hedge but difficult to propagate. *Justicia gendarussa* can be clipped to 15cm and allowed to grow upto a metre. *Lagerstroemia indica* in many colour variations, though bare in the cold weather, is a fine flowering hedge in April-June and if two colours are planted alternately provides a pretty effect when in bloom.

Ligustrum neilgherriensis, one of the privets, is not to be despised and the *Lantana,* though so frightful a weed, can be retained as a dwarf plant with frequent clipping though the non-seeding types are recommended. *Malyaviscus* can be cut back to form a 1 metre hedge or allowed to grow 2-3 metres high. When in bloom it is very effective.

Murraya (Kamini) and Tropis aspera (Sheora) are slow growing but clip well. Jasminum pubescens, Coffea bengalensis, Barleria (Jhati), Ixora especially I. ragoosula, I. coccinea and I. strica. Daedalacanthus and Strobilanthes are flowering hedges that are well known and easily grown. Roses form excellent hedges and many of the cacti such as *Cereus, Opuntia* as well as *Euphorbia antiquorum, E. trigona and E. tirucalli* are used in exposed positions for hedge work. *Pedilanthus tithymaloides* will be found as hedges surrounding village huts. Being poisonous to cattle it is left severely alone.

Sesbania aegyptiaca (Jaint) is a fast-growing hedge reaching a height of 2 metres in a couple of months; it must be grown from seed and is a great favourite in the drier parts of India for a tall temporary boundary hedge. After two years it commences to become rugged and should be replaced. Another quick hedge is *Cajanus indicus* (Arhar) but this grows too rapidly in Bengal.

Thuja orientalis is an excellent evergreen hedge, 2-3 metres in height. At an elevation of over 1000 metres a different type of plant is cultivated for hedges. Here are some varieties: *Spirea, Kerria japonica, Berberis* (Barberry), *Habrothamnus elegans, Fuchsia, Heliotrope, Euonymus, Ligustrum* (Privet), *Buxus* (Box) *Hydrangea, Philadelphus, Hibiscus syriacus, Cupressus* and *Ancuba.*

Trees: For very tall hedge several trees lend themselves to pruning: Polyalthia longifolia is one of the best, *Putranjiva roxburghii, P. acuminata, Anogeissus, Diospyros embryopteris Grevillea robusta* and *Pithecolobium* are others that can be recommended *Pongamia glabra* and Inga dullics are also used. *Thevetia nerifolia, Erythrina* of many kinds, *Haematoxylon campechianum, Schinus terebinthifolius, Cryptomeria, Cupressus* and other conifers are considered good when cut back to a height of 3-4 metres.

Casuarina equistifola if pruned before large trunk forms, makes a very neat hedge. Acacia farnesiana and Parkinsonia aculeata are both useful but do better in dry situations. *Pisonia alba* (tree lettuce) which forms hedges of pale coloured foliage, is grown in South India and Ceylon but not elsewhere to any extent. Commence cutting back when the seedlings are 20 cm high and repeat when the shoots are 40 cm a quick thick hedge from the very beginning is then ensured. *Thevetia nerifolia* is poisonous to cattle and goats and make a neat-clipped hedge 3-4 metres high.

Climbers: On a wire fencing Jacquemontia violacea is ideal, Ipomea palmata (Railway Creeper) and I. sincrata and Antigonon (Sandwich Island Creeper), Clerodendron splendens, Ficus repens, Passiflora pruinosa. Tristellatia austrulis, Thunbergia fragrans and T. alata and Vallaris hynei are good for screening. For a very shady situation Rhycospermum jasminoides, Asparagus plumosus and Lygodium scandens are suggested.

Palms: *Ptychosperma mcArthuri* will grow 4-6 metres and not spread more than 1.5 to 2.0 m in width. *Rhapis flabelliformis*, on the other hand, while growing form 2-3 m in height, sends out runners, which must be constantly removed. *Areca lutescens* grows as tall as the *Kentia* but bushes out more.

Planting Distance: This depends on the size of the plant. Trees should be planted 2-3 m apart, shrubs 15-50 cm apart and *Alternanthera*, etc., (edging) 2 cm apart if planted in a single row but when staggered double the distances. *Duranta, Lawsonia* etc. can be thickened when the hedge has grown up, by having the pruned cuttings, planted in between or on either side of the growing bushes. For hedge planting calculate 9 plants per mete row of planted singly 10 cm apart and 12 in a double row if planted 15 cm apart staggered.

***Care of Hedges*:** The first requirement for a hedge is impenetrability and the next neatness. Any prickly plant, such as *Inga dulcis* or the thorny type of *Duranta*, will not ensure an absolute cattle or thief proof hedge in its earlier stages, so run a barbed wire through the bushes form the ground or plant a row of pineapple, *Agave*, prickly cactus or thorny *Euphorbia* at the base and the case is altered. Always keep jungle away from the roots of a hedge. In the dry season fork up the soil and flood' when Mendhi (*Lawsonia*) or *Duranta* are pruned, during the wet weather, plant the cuttings so as to fill gaps.

Fertilisers: Usually hedges receive little attention till they become thin. The plants are pruned heavily in the rains, add manure at the close of the Monsoon. Feed hedges, especially those that are clipped regularly by giving superphosphate of lime and bone meal 50 gms each, 12 gms sulphate of potash and 50 gms sulphate of ammonia per two metres row, every six months. Once a year dig in cow manure as the rains break. This might seem superfluous but the results will amply repay the attention given

***Pruning of Hedges*:** Some varieties have to be left to grow as in nature, *i.e. Pandanus, Euphorbia*, etc. but kept within bounds by the removal of side growth or the central head. In *Euphorbia* and *Opuntia*, the taller portions broken off can be replanted in the gaps to thicken the hedge. For low hedges, 50 cm to 1.0 metre high, clip frequently during the rains but once a year cut back the old wood below the maximum height to restart a new top do this in May or June.

Use a sharp pair of pruning shears for small twigs but a secateur will be necessary for larger stems. It is best to stick to a simple shape instead of attempting battlements, turrets, balls, columns, etc. arches can be grown on an iron rod bent to shape, the shrub kept in position by being loosely tied with strong wire to the rod.

Shapes of Hedges: A hedge is apt to become wedge-shaped the top widening out, if not carefully clipped square. A rounded top is difficult to keep unless the type of hedge is slow, compact growing and small leaved. Hedges are also cut on a slope but here too the Mali often makes a mess of the shape. It is best to have wooden or iron stakes painted green, sunk in the hedge at intervals on which string can be tied and used as a guide. Stick to a square cut hedge.

Topiary: Topiary is practised only in few traditional gardens. The plants that have small foliage and are of slow growth are good for using in topiary. The results required takes years to attain and one careless clipping will spoil the entire design. Stating with a ready-grown plant, stems are bent and held in position with wire if they do not naturally grow at the required angle. A rough clipping to get the general outline follows and as new shoots develop these are either trained to fit in with the design or removed. Simple shapes, a ball, spiral, table, cube etc., are not so difficult to obtain, it is when birds or beasts or even man is to be moulded that the topiarist tackles a tough proposition. Topiary work gives an old world appearance to a garden but these clipped bushes should not be scattered aimlessly all over the garden. A formal garden is best suited to topiary work. For simple shapes the following can be used *Hibiscus,* Thuja and Cupressus, Putranjiva and Polyalthia but for the more intricate Murraya exotica, Duranta repens, Ehretia buxifolia and Tropis aspera are used. When using wire to model the bus hor tree to the rough outline care should be taken that this does not cut too deeply into the bark. Stoppage of sap will mean poor growth or death of a stem.

Ornamental Foliage Plants

The shrubs are grown in the garden for handsome and richly variegated foliage *e.g.* Acalypha tricolour, Nandina domestica, Codium variegata, Manihot variegata, etc.

Some shrubs have handsome foliage and produce good flowers too. They add to beauty in the garden tremendously *e.g.* Bougainvillea, Buddleia asiatica, Buddleia madagascariensis, Hamelia patens, etc.

***Propagation of Shrubs*:** The shrubs are propagated by seeds, cuttings and layering.

By Seeds: Seeds are collected when they are fully ripe. They are dried in the shade and stored in airtight bottles in dry places. In rainy season, nursery is prepared and seeds are sown. Seedlings are transplanted in prepared ground. The shrubs which are mainly propagated by seeds are: Stenolobium stans, Thevetia peruviana, Caesalpinia pulcherrima, Calliandra sp. etc.

By Cuttings: Majorities of the foliage shrubs are propagated by cuttings. The best season for propagation by cuttings is rainy season, although cuttings can be made in February-March if water supply is not limited. The cuttings are made 15-20cm long and planted in pots or beds containing ample of sand. The plants raised from cuttings are planted in the next planting season. The shrubs propagated by cuttings are Hibiscus rosa-sinensis, Achania malvaviscus, Jasminum sambac, Hamelia patens, Bougainvillea, Cestrum diurnum, Cestrum nocturnum, etc.

By Layering: The shrubs which are not easily propagated by seeds or cuttings are propagated by this method. In rainy season, layering is dome by removing the bark of 2.5-3.0cm long and wrapping by garden soil and sand mixture or sphagnum moss. The shrubs like Bougainvillea, Ixora sp. Jasminum sambac and Jasminum multiflorum (ground layering) etc. are propagated by layering.

Acalypha (Euphorbiaceae)

It is native of East India. This shrub is famous for its copper coloured and variegated foliage, which is highly desirable for contrast effect in landscaping of garden, buildings etc. It is a medium shrub growing upto 1.2 - 1.5m prefers semi-shady situation flowers are red and long like catkin. This herb produces the flowers throughout the year but profusely from February to July, and is very easily propagated by cuttings.

Achania Malvaviscus (Malvaceae)

The foliage is dull green but flowers are bright scarlet, pink half-open, producing throughout the year. Easily propagated by semi-hardwood or hardwood cuttings.

Hamelia Patens (Rubiaceae)

It is native of tropical and sub-tropical America. This shrub is very famous amongst gardeners for attractive foliage and scarlet flowers, which are produced throughout the year. It is a tall shrub growing about 2.4-3.0m high. Due to profuse branching it shapes well appears elegant. Foliage is green and greenish bronze, which stands clippings and trimming to any shape. This shrub can serve very well either as standard, specimen or hedge in the garden. It can be very easily propagated by cuttings.

Iresine Lindenii: Grows upto 11/2′ to 2′ even having rose red coloured foliage. There are tow distinct species of great value. Suitable edging for large plants in the garden.

Pilea Muscosa: There are a few varieties with spreading habit. Useful as covering plants in shade for edging rockeries etc. Thrives in shady and semi-shady conditions. Grows from 3′ to 8′. It has fern like foliage and the plants are succulent in nature.

Peltophorum Ferrugineum: A quick growing tree with fine feathery foliage. The tree gives a good show and shade. The flowers are borne in panicles. Brown spots in clusters, which adhere to the branches for quite long, time and are very attractive. Propagated by seeds. Good for avenue.

Callistemon Lanceolatus: It is a small erect growing tree with a near habit of growth. The foliage is attractive which consist of narrow stiff, lanceolate leaves and beauty of its blossoms. It is very beautiful in April with its bottlebrush like crowded cylindrical spikes of brilliant crimson scarlet flowers. Propagated by seed and air layerage.

Sweetheart Plant: Philodendron scandens, sometimes called P. Oxycardium, is one of the more popular houseplants. It comes from the tropical jungles of the South and the Central America. It has heart-shaped deep green leaves and is a rapid climber sticking on supports with roots produced from the nodes. Leaves of juvenile plants reach 10 to 15 cms in length on six to eight cm long stalks. Leaves of mature plans are larger. The leaves, which have acutely pointed tips, look transparent when they are new, but quickly become deep green as they grow to maturity. Regular pinching out of the growing tips makes the plant bushy. The climber can be planted in window boxes, to cover trellises and as a cascading plant in hanging pots provided the plant is not in direct sunlight. Being a jungle plant, it needs good watering and frequent misting or wetting with clear water. Philodendron scandens loves to grow in a leaf mould based friable soil. Being a fast grower, the trailing stems of the plant should be occasionally turned back and tied on the supports. As the plant roots at the nodes, propagation is easily done by keeping two or three leaved cuttings in sand or loose soil.

Succulents

Succulence in plants occurs due to lack of adequate supply of moisture together with heat and sunshine. It is necessary for the plants on high mountains to withstand drought and vegetation of succulent appearance are commonly seen. In the seacoast or salt deserts salinity results in the formation of succulent plants. Most of the succulents occur in the South African deserts and in the Cape Province. Succulents also occur in the deserts of the Sahara. Various forms of giant succulent plants are native to the Isle of Socotra. Many beautiful and interesting succulents come from Canary Isles. In the stony and the calcarious soil of he Southern and the Western sides, numerous succulents such as, Agave, Echevaria, Sedum and xerophyllum thorny bushes grow in abundance.

Succulents are among the most specialized plant form and they are the expression of peculiar condition of soil and climate. As the plants grow in the severe shortage of water in the deserts and semideserts they are adopted to conserve water and to reduce transpiration. The water storage tissue occupies almost the whole leaf or the shoot and is surrounded by a thin layer of assimilating tissues. The leaves or the shoots may contain water as much as 95% of their volume. The water content in the leaves is very mucilagenous, which is very important for its retention of water. During the dry period, the plants are usually completely at rest and look dead since they have shrunken and the surface of the leaf dried up. But they recover very quickly after the first rain, in a short time produce new leaves and show new growth by the swelling of the plant body. In ornamental horticulture, these succulent plants are used for their peculiar shape, decorative foliage or showy flowers and constitute the most important plants in the rock gardens.

***Soil and Manures*:** Light porous soil is best for succulents. Undecomposed organic matter should never be used. Well-decomposed leaf-mould and cow manure, loamy soil, sharp sand, crushed, brick and charcoal constitute the ingredient of the compost for succulents. For many species, addition of lime is recommended, the best form being mortar rubble. For most of the Crassulas the following compost is recommended: Leaf-mould 3 parts, well-rotted cow manure 1 part, Loam 1 part, Crushed brick 1 part, Sharp sand 6 parts. For Euphorbia, Aloe and Agave use a compost consisting of 1 part each of cow manure, Leaf-mould and loam and 2 parts sharp sand; Echeveria, Aeonium and similar less sensitive plants generally require loamy soil which is not too prorous.

Propagation

i. *Propagation by Seeds:* Since almost all succulents flower and produce seeds, propagation by means of viable seeds is a practical method. Raising from seed becomes essential when large number of plants of one species are to be grown for commercial purpose. The best time for sowing seed is the spring. At a temperature between 30 to 35°C and the relative moisture content of the air of about 90% most seeds germinate quickly. After germination of seeds the seedlings should be removed in drier place. The seeds compost should also be very porous mainly containing well-rotted dry leaf-mould mixed with twice the quantity of washed coarse sand. Shallow pans are commonly used for sowing seeds filled with the compost and covered with a thin layer of sand. The seeds are then scattered thinly on the sand again covered with sand to a depth not greater than the size of the seed itself. After sowing, the pots should be kept in glass frame or covered with a glass pane. Watering is done either by spraying through a fine rose or keeping the pot on a vessel containing water. When the seedlings come up, the young plants are transplanted in a compost cotaining 3 parts Leaf-mould, 1 part old cow manure mixed with soil, 1 part mortar rubble, 1 part loam, 6 parts clean sand. The young seedlings should be kept dry during the resting period. Like older plants the young seedlings, however, should not be exposed to extreme climatic conditions.

ii. *Vegetative Propagation:* Almost all succulents can also be propagated vegetatively. As the succulents are fleshy, the parts removed must be carefully handled. The growth is cut off with a sharp knife and kept in an airy, dry place so that the cut surface becomes dry and covers itself with a shine. Cuttings should be taken from mature growth, little woody rather than too soft, for soft cuttings rot easily. It is not always necessary to use the terminal cuttings, lateral shoots may also be used. Aloe and Agave produce offsets and these are used for multiplication of plants. When large enough, these offsets are separated from the plants, a little below the surface of the soil and treated as cuttings. In many cases offsets are found to produce roots while attached to the plants and they can be potted after separation. Though it is not possible to multiply all types of succulents from leaf cuttings, species of Crassulaceae lend themselves successfully to this method of propagation.

The leaves are carefully detached from Crassula, Cotyledon, Echeveria, Sedum etc. And dried like the stem cuttings. Haworthia and Gasteria also produce shoots at the base of the cuttings. A few succulents such as Kalanchoe bears on the edges of the leaves small adventious buds which may be detached and used as cuttings. The rooting medium for all types of cuttings whether from shoot, leaf or buds should contain sand and dry Leaf-mould in the proportion of 2:1. As a rule shoot cuttings root quicker than leaf cutting and the rooted cutting should be carefully removed and potted in the compost recommended for the plant.

Watering

Watering of succulent plant is one of the most important points. Hawarthia, Gasteria should not be allowed to dry too much and the soil should be only moderately moist. During the winter month, the soil for Huernia and Stapelia should be kept just moist so that the roots are not dried up. Euphorbia should be watered freely during the growing period but very little during the resting period in the winter months.

When watering is done the soil should be soaked thoroughly and sprinkling of water on the surface soil be altogether avoided. Whenever possible rain water should be used in pots.

Repotting

Like cacti the resting period of most of the succulent comes in winter and the growing period during the spring and the summer months. This is, however, not true with all types of succulents as some species begin to grow in June, July and August and their resting period is in the spring and the early summer. Like other plants succulents should be grown in clay pots which are porous, *i.e.*, water and air pass through them. The pots must be provided with drainage holes so that superfluous water can pass out readily. The shape and size of the pots should be in proportion to the root system of the plant. Aloe, Agave etc. should be grown in large posts on account of their large roots. Many Euphorbias also need deep pots. For shallow-rooted plants like most of the Gasteria, Haworthia, Stapelia shallow pots should be used.

Repotting should be carried out at the beginning of the growing period. Quick-growing plants should be repotted once every year. Species which grown less freely need not be repotted every year. The plants should be taken out very carefully with minimum injury to the roots and then firmly replanted in a bigger sized pot in the compost of the same composition.

Common Succulents

Adenium Obesum : Fleshy, thick-twisted stem grown in rock garden, about 2 m high; flowers pink.

Aeonium: Succulent herbaceous plants with woody and branched stems, leaves in the form of rosette at the ends of the branches. Some species have no distinct stem and rosette arises close to the ground.

- A.haworthii has bluish green leaves, edges reddish brown and look like flower of marble.

Agave: These are the common succulent plants with xerophytic appearance, leaves in the form of rosette, firm, hard and sword-shaped, usually spiny at the margin and tip, sometimes smooth at the margin but the apex is stout and pointed. The plants die after flowering. Adventitious buds often form at the axils of the panicle branches.

- A.americana—Stemless rosette, margins spiny, curved. The variegated forms are commonly grown in rock gardens.
- americana var. Marginata. Variegated form with margin pale yellow or cream.
- americana var. Mediopicta—Leaves with yellowish stripe in the centre.
- A.angustifola var. Variegata—Compact rosette, very firm leaves, margin whitish, middle green.

Aloes are a group of succulent plants, hardy and widely found in tropical countries, plants stemless or with short stem or sometimes tree-like. Leaves mostly in rosette, fleshy, margin entire or toothed surface may be spinous sometimes with markings or stripes.

- Aloe variegata—Plants form densely rosette of leaves arranged in three ranks, upper surface concave, lower surface keeled, dark green with transverse white bands arranged irregularly.
- A.zebrina—Leaves linear-lanceolate, in rosette, upper surface flat fleshy, green with numerous white spot.
- Caralluma—Succulent perennial herbs, 4 to 6 angled dentate sometimes creeping stem or semi erect grow in clump.
- C.sprengeri—Stem much branched 8-10 cm high, 4-angled, angles, toothed, light green with reddish spots.
- C.turneri—Plant taller, stem 4 angled dentate, green with purple spots.

- Ceropegia—Succulent perennial harb, stem leafless or twining with opposite leaves, lanceolate or cordate. Flowers showing various colours.
- Cissus cactiformis—Climbing succulent plants 4-angled, winged, constricted at the internode. Cissus quadrangularis is also a climbing shrub, 4-angled constricted at the nodes, leaves cordate, falling off.
- Cotyledon—Succulent herb or shrub, usually grows in clumps. Leaves spirally crowded, thick and fleshy. Flowers pendulous from terminal branches.
- C.dobiculate—Erect shrub, leaves opposite crowded at the tips of the branches, obovate, narrowed into a short stalk.
- Crassula—This genus has more than 300 species and forms the most important group of succulents. These are usually herbaceous plants or semishrub with succulent branches and twigs. Leaves, opposite often crowded into a rosette usually sessile. Flowers small, mostly numerous in terminal or lateral cymes.
- Echeveria—Perennial herbaceous much branched succulents. Leaves spirally arranged in rosette, fleshy and thick, various shapes and size. Flowers appear from the leaf axils, usually bell-shaped.
- Echeveria agavoidis—looks like a rosette of leaves, compact ovale with brown tips. Flowers pink.
- E. glauca—This species has rosette of leaves, rarely any distinct stem with many offsets. Leaves broadly obovate, spatulate, wedge-shaped towards the base, margin reddish.
- E. elegans—It also forms plants with rosette of leaves, obovate rounded above, pointed tip.

Euphorbia: Euphorbias are also common succulents in tropical country. They are usually thorny, fleshy. In many cases they resemble cacti in many ways. Few species have large prominent leaves.

- E. antiquorum—Much branched thorny succulent, stem 4 to 5 angled, branches usually 3-angled constricted joint-like. Spines spreading on small roundish shields.
- E. candelabrum—Tall tree-like branches 4-angled.
- E. coope—Spiny much branched succulent, 5 to 6 angled, divided into many joints usually constricted, angles with grey throny edges and spines.
- E. tirucalli—Much branched, woody succulent, branches jointed, dark green in colour.

Fourcroya: Plants are like agave with tufted leaves, margin spiny or dentate leathery.

- F. gigantea-Large oblong oblanceolate leaves, flat undulate margin, fresh glossy green, margin entire.

- F. giantea var. Mediopicta—leaves with creamy white or pale yellow longitudinal centre.
- F. selloa var. Marginata—leaves variegated, margins yellow.

Gasteria: It is a showy succulent, leaves long, thick firm mostly distichous.

- G.marmorata—leaves lanceolate long, densely distichous, rounded tapering apex, dark green, darker mottled.
- G.parvifolia—Stemless, distichous leaves, reddish, glabrous, glossy spotted on both sides.

Graptopetalum: Perennial plants, look more or less like Echeveria. Leaves in rosette, thick and fleshy.

- G. rusby—Plants small, rosette of compact leaves, thick ovate, whitish.

Haworthia: Low-growing perennial herbs, leaves in rosette or in compact rows, fleshy with smooth tubercles, tip usually stiff.

- H. attenuata—This species has few sub-species.
- H. attenuata var. Clariperla—Leaves in the form of rosette, wholly covered with small tubercles arranged in series or irregularly scattered.
- H. greenii—It bears large number of leaves, reddish or light brownish green in the sun. Leaves compressed, loosely arranged incurved; upper surface glabrous, back side with longitudinal lines.
- H. reinwardtii—It is popular species of Haworthia with few sub-species. This species usually has large number of leaves loosely arranged, tubercles conspicuous arranged in various fashions.

Kalanchoe Tubiflora: Erect plant, mostly unbranched, leaves almost cylindrical, somewhat broad in the upper surface.

Nolina : Small tree, succulent stout trunk swollen at the base and tapering towards apex. Leaves long, narrow, leathery.

- N.recurvata—Base of the stem almost globose. Leaves thin, long recurved, tuft.

Pachyveria: Intergeneric hybrid between Pachyphytum and Echeveria. Plants look like Echeveria and Pachyphytum. Leaves succulent, fleshy in rosette on short stem.

Pedilanthus: These are succulent shrub or sub-shrub with latex.

- P.carinatus—Much branched shrub, dark green erect stem. Leaves ovate-lanceolate, leathery, fleshy

Pelargonium: Several species of Pelargonium commonly grown in milder climate have succulent stems. P. juttae has globular stem.

Portulacaria afra-Succulent shrub; stem horizontal spreading. Leaves opposite, obovate, thick, glabrous, shining green.

Sedum: These are annual, biennial or perennial herbs or shrubs. Branches herbaceous or slightly woody, dichotomously branched erect creeping. Leaves mostly alternate, flat or cylindrical and in most cases they are closely set on the stem.

- S.compactum—Low growing herbaceous plant. Leaves obovate, obtuse on short rosette.
- S.morgaianum— Creeping sub-shrubs. Leaves numerous, thick fleshy, lanceolate
- S.neudum— Small sub shrubs. Leaves alternate, obovateoblong, thick, blunt at the tip.

Sempervivum: These are a group of low-growing perennial herbaceous plant found in the rock gardens of temperate countries. The leaves are arranged in the form of rosette, thick fleshy, green, reddish or bluish in colour, glabrous or hairy.

Senecio: This genus comprises 1300 species and distributed all over the world.

Stapelia: Perennial succulent herb, stem fleshy erect, branched at base and forming clumps, 4-angled, the angles are toothed, green and also reddish. Flowers are of curious shape usually arise from the base of the stem.

Cacti

Cacti are a group of plants with peculiar shape and size and mostly adopted for desert life. The adverse environment has made these plants unusual in appearance and provided morphological, anatomical, and physiological safeguards to thrive and flourish under such condition. The Cactaceae are perennial herbs or shrubs, annual species being non-existent, with more or less fleshy stems, usually jointed. The leaves are rarely perfect and permanent, but are generally absent or rudimentary or fugacious or reduced to mere minute scales. In Pereskia the leaves look like normal flattened leaves as in other dicots.

In the genera Echinopsis, Rebutia, Lobivia, Echinocactus, Astrophytum and Lophophora, the stem is globular or avoid or cylindrical, very fleshy, usually with many ribs, acute or rounded, straight, wavy or spiral, more or less notched, with areoles and spines and the flowers are borne on the areoles at the top or at the sides.

The stem of Opuntia is always a typical joint whether it is globular, cylindrical or flattened. The joint is, therefore, a stem which when properly formed ceases to grow at the apex, but continues to grow in size and thickness until it reaches its full development and then the rudimentary fleshy cylindrical leaves drop off and the spines grow to their full size.

Spines are of two types. The real spines, vary in length according to species and in most cases very stiff, acute, needlelike, awlshaped, tortuous, arched or hooked and develop on the outer or lower areole or bud. On the upper or inner areole, just above the spines hairs develop which are flexible or more or less stiff bristles. The spines and bristles of cacti are often a terror particularly in the operations of planting and repotting.

The largest and finest flowers of the Cactaceae are mostly nocturnal. They open at sunset and display all their glory in the dark hours of the night and droop and fade away at dawn.

Climate

Cacti can tolerate wide range of temperature. Very high temperature or very low temperature adversely affects the growth of many species of cacti. At a temperature above 35^{0}C under the blazing sun, plants may show sign of scorching or turn yellow. In such condition the plants should be shaded off. Most epiphytic species require half-shade and a moist atmosphere.

Soil and Manures & Fertilisers

The soil for cacti should be porous and permeable. Almost all species thrive well in a porous calcareous soil. Lime mortar from old buildings, hard limestones, marble or even crushed valves of oysters or other shell fish, roughly powdered and passed through a sieve to remove the fine dust, provide an excellent material to add to the soil. Some old Leaf-mould, thoroughly rotted should form a part of the compost. Very well-rotted cow manure is also an useful ingredient in the compost for the cacti. Undecomposed organic matter is sure to give trouble and causes rotting of roots. Chemical fertilizers force growth and due to excessive succulence the plant becomes susceptible to drought and infection.

Small quantities of bone meal, superphosphate or basic slag may be added to the compost. The compost should be watered and allowed to remain for few months before using it. A small quantity of charcoal powder also helps to keep the soil porous and sweet and prevents infection of roots. On the other hand, a soil, which keeps moist or forms a green coat or crust on the surface, is either badly drained or unsuitable and should be changed without delay.

Planting and Potting of Cacti

Many hardy cacti are also grown in ground particularly in the rockeries. The land should be sloping and the soil porous and calcareous. Garden loam 2 parts, sand, old mortar or limestone gravel and Leaf-mould 1 part of each and some crushed charcoal are suitable for Cleistocactus, Expostoa, Oreocereus, Borzicactus etc. For Cereus, Cephalocereus, Pachycereus, Lemaireocereus, Harrisia and other Cereus, use garden loam 2 parts, old mortar or limestone gravel, Leaf-mould or manure 1 part of each: some crushed charcoal may be added. The best time for planting and potting is just at the commencement of the growing season in spring.

The soil should not be too wet and it is always safe to place some dry soil beneath and around the roots. If roots are damaged at the time of planting, watering should be stopped for few days to allow the wounded roots to dry and heal. In order to promote new roots little watering is done to keep the compost just moist. Water may be given more freely later on, when the temperature rises and new growth develops.

Well-baked earthen pots are suitable for cacti, glazed or painted pots hamper the aeration of the roots. In potting or repotting cacti, it is safe to select the pots just large enough to contain the plant. Pots which are too large have two disadvantages. When watered they retain too much moisture, retain it too long and appear as constant danger to the plant. Moreover, the roots take too long to reach the sides of the pot and spread, thus causing a delay in growth and flowering.

The pots should be filed for about one-third with crocks or other drainage material, the large curved pieces to be placed on the holes and bottom and the hole topped over by a thin layer of thick gravel, shingle or crocks, broken into small pieces.

If it is found that the roots have become pot-bound it is advisable to break off the pot with a sharp blow, instead of risking damage to the plant in trying to take it out and save the pot. The plant to be repotted should be kept dry so that it easily comes out without injuring the roots and crocks are removed carefully. The plant should be replanted just as deep down as in the previous pot. Unless the soil is in a perfectly dry condition, newly potted lightly sprayed once or twice a day and placed in shade for about one week, afterwards in half-shade for at least an equal period before removing them to full sun.

Watering

It is very important practice for cultivation of cacti. Under natural condition these plants are adapted to server drought. Their external structures and also the physiological condition clearly indicate very little requirement of water.

In the summer months, when the cacti grow, watering is done frequently. If these plants are kept in open in small pots daily watering may be necessary. Every time the soil is to be drenched completely so that excess water comes out through the drainage hole. At the resting period, during the winter months roots remain inactive and watering should be minimised gradually from the autumn. Water should not accumulate in the depression of the plant. In Echinopsis, Echinocactus, Rebutia the growing point on the top may rot. Fine spraying or syringing of water may be done in the morning.

Propagation of Cacti

Propagation of cacti is done by both seeds and vegetative methods.

1. *Sowing of Seed:* Cacti can be grown from seeds but the seedlings grow slowly and take long time to form a specimen. It is commonly used by commercial growers and amateurs who specialise and aim at a large collection of these plants.

   The seed compost consists of equal quantity of leaf-mould and sand and some amount of loam, powdered brick and char coal. Seed pan or shallow pans are used for sowing and half the depth of the pot is crocked. The pans are filled with the compost leaving about 3 to 4 cms on the top. The soil is then gently pressed and levelled. Seeds are sown uniformly, pressed down in the soil and thin layer of sand spread on the top. The compost is watered by placing the pot in water and allowing it to enter through the bottom and side holes. The pot is covered with a paper or glass pane. Cacti can be sown successfully from March to September. Germination may take place from 7 days to one month. When the sedlings are 1.5 to 2 cm in diameter and in the case of tall-growing species unless about 4 to 5 cm high, they are taken out without injuring rootlets and planted in small pots containing the same compost is shade and watering is best done by spraying after a day or so. A number of seedlings may also be transplanted into a bigger pot at a distance of 3-4 cm.

2. *Propagation by Cutting or Offset:* Propagation by cutting of offset is an easy method of getting large plants and also the only way of perpetuating hybrids and valuable plants. Species with solitary globular stem cannot be propagated by this method. Tall-growing species with single stem sends out shoots if the top portion is cut which can be used as the cutting but the specimen plant is lost. The branches thus formed are also detached and multiplied.

   The cuttings are cut clean with a sharp knife and kept in airy and shady place for drying. The period of drying may vary from seven days to one month. The cutting is then planted in sand and leaf-mould or in the usual compost and kept in semishade till rooting. During the time of root formation watering is done sparingly even in the summer months. Formation of roots in the cuttings is indicated by new growth on them.

   Cutting of Cerei including Harissia, Selenicereus should be of covenient length. Pachycereus, Lemaireocereus are cut at a constriction. In branching species of Mammillaria and Rebutia, the clumps are divided into smaller clumps. Species of Trichocereus, Cleistocactus, Echinocereus, etc. are best propagated by separating the suckers which arise from the lower part of the stem. Cuttings of Epiphyllum are best taken across the broad part of the flat stem.

3. *Propagation by Grafting:* Propagation by grafting is a useful method of propagation. Generally a weak growing species is grafted on a strong growing species is grafted on a strong growing stock, the weak scion improves the vigour and stock is weakened. Species of Selenicereus, Trichocereus, Harrisia, Cereus, Opuntia, Cleistocactus, etc. are commonly used as stock.

The grafting may take the form of an inverted wedge. The top of the stock is cut as a wedge which is inserted in wedge shaped cavity at the lower end of the scion for proper union. It is very important that the sides of the wedge and cavity should come in close contact and the top of the stock fit in the bottom of the cleft.

For grafting Cephalocereus, Oreocereus and Espostoa, clean horizontal cut is made on both stock and scion. The scion is kept fixed on the stock by giving a moderate pressure by tying the scion with the potted stock. To avoid injury a piece of folded cloth is placed on the top of the scion and the string is tied on it. After a month the pressure is released. The cut end of the stock and scion instead of being horizontal may also be inclined at an angle of 45 degrees as in the case of tongue graft.

Cacti can also be grafted by making a clean cone or inverted pyramid at the base of the scion and inserted in a caving made into the upper end of the stock. The scion is then tied in the usual manner. Grafting is also done in the summer months when the humidity is very low. High humidity causes rotting.

Common Cacti

i. Astrophytum is known as Starcactus because of the large well-formed ribs. Surface is covered with star like hair or scales. Stem globular, flowers borne on the aeroles near the top. Astrophytum asterias, A. myriostigma and A. capricorne and their varieties are very popular.
ii. Borzicactus has slender, weak stem, many ribs, with numerous aeroles. Cleistocactus also appears like Borzicactus, stem slender, erect, often throwing off shoots or branched with numerous ribs.
iii. Cereus forms a group of all erect of prostrate, branched, ribbed or having angular stems. Cephalocereus has tall columner stem solitary or branched at the base. Aeorles woolly or hairy. Lemairocereus has tall erect stem branching from the base. Aeroles mostly with stout spines.
iv. Oreocereus: columnar in appearance, hairy, often known as the Old Man of the Andes. Spines are very strong. Red flower appear on matured plants. O. celsianus and O. hendrick senianus form good specimens.
v. Echinocactus has large barrel-shaped appearance with prominent ribs and numerous strong species. Large yellow flowers appear at the top. E. grusonii

is a popular golden yellowed spined cactus. The stem may be 70 cm across and about 80 cm high. Echinocereus is a group of hardy cacti, very easy to grow. Stems roundish, oval or elongated, branching at base, erect or prostrate, ribs many. Aeroles woolly with spines.

vi. Echinopsis is another group of hardy and widely grown cactus. Stem cylindrical, globular, solitary or in cluster with many acute ribs continuous or notched. At least 36 species are in cultivation.

vii. Epiphyllum is also an epiphytic plant with woody main stems and flat leaf like branches, toothed or notched on both sides. Aaeroles small situated in the notches often associated with bristles. Flowers large, showy open at night time. Epiphyllum hookeri, E. phyllanthus and E. stricutum are common: large number of hybrids are cultivated in the western countries.

viii. Espostoa: Woolly or hairy columnar plants, spines are less conspicuous than Oreocereus. Small flowers are almost hidden. E. lanata and E. dautwitzii are commonly grown by amatuers.

ix. Ferocactus means ferocious cactus and the spines are long, thick and stiff. Stem is large round or cylindrical with permanent ribs.

x. Gymnocalycium has about 50 species with globular stem and many well-marked ribs straight or irregular. Spines prominent; flowers large.

xi. Mammillaria is the largest genus in Cactaceae with 200 recognised species. Plants are globular, oval or cylindrical branched or in cluster. Ribs are absent, tubercles in spiral rows, woolly or hairy in the axils. Flowers small, borne in the axils of the tubercles around the upper part of the stem.

xii. Notocactus: Globular or columnar in shape easy-growing and free-flowering. The flowers are usually yellow with red stigma. N. concinnus, N. haselbergii, and N. leninghausii are few widely grown species.

xiii. Opuntia is a very commonly grown cactus, naturally occuring on road sides, rocky or sandy places. They are differentiated by their flat-jointed stem. Some are quite attractive, having red or yellow bristles. The genus includes over 300 species.

xiv. Parodia: Small plants, mostly globular, some are cylindrical, very spiny; bears large attractive flowers. P. aureispina, P. chrysacanthion, P. microsperma form well-shaped plants and bear colourful flowers.

Rebutia has many species of small globular plants producing offsets. The flowers appear on small plants and vary in colour R. deminuta, R. fiebrigii, R. marsoneri, R. senilis are popular.

Bonsai

Bonsai is an old Chinese art of growing trees. Bonsai comprises a tree or shrub planted in a small container for developing as a miniature plant showing the general appearance of that plant species found in nature. It differs from a pot plant where the foliage and flowers are important whereas for Bonsai the appearance of the plant in a miniature form is to be maintained for many years.

Plants Suitable for Bonsai

Although the suitability of plants for developing as Bonsai has been tested in Japan consisting mostly of sub-tropical and temperate plants, very little information is available on the response to the growth of tropical trees in miniature form. Late Shri V. P. Agnihotri has developed a technique of dwarfing trees commonly found in this country. The suitability of a tree to develop as a Bonsai depends on various factors:

1. The plant should be hardy so that it can be grown in a small container for many years with all the manifestations of a living plants.
2. The trunk should develop a natural appearance.
3. The branches should grow in natural but artistic forms.
4. The growth of the plant and its appearance should be harmonious with the shape of the container.
5. The miniature plant showing seasonal variation in growth and flowering is a very interesting feature of Bonsai.
6. Plants of low height and strong trunk, thick at the base are good as Bonsai.

Among the tropical trees that thrive well as miniature plants are: Adonsonta digitata, Anthocephalus cadomba, Bombax malabaricu, Adenanthera pavonina, Brassia actinophylla, Butea frondosa, Caesalpinia coriaria, Erythrina cristagalli, E. parcelli, Ficus religiosa, F. bengalensis, Jacaranda mimosaefolia, Kigelia pinnata, Putranjiva roxburghii, Tabebuia chrysantha and Thespesia populnea.

Several tall shrubs like Adenium obesum, Brya ebenus, Fortunella japonica, Hamelia patens, Hibiscus schizopetalous, Jatropha podagrica and Murraya exotica form very attractive dwarf plants. Woody climbers like Combretum, Derris scandens, Hiptage madhablata, Roupellia grata and Jasminum auriculatum can also be trained as Bonsai. Among the conifers grown in tropical conditions, Juniperous prostrata and Pinus khasiana will form good Bonsai.

Growing the Bonsai

Plants suitable for growing as Bonsai are planted in small containers. In tropical climate, the monsoon is the best season for planting or transplanting. For making

Bonsai, plants growing wild or seedlings grown in the nursery for several years should be carefully uprooted, grown in ordinary pots for a year or two and then planted in a shallow container. Plants may also be obtained from layering of a large branch, grafting or by raising seedlings in a pot.

It is advisable to grow the plant in the ground for a year or two which helps in developing a strong root system and healthy branches. They are the transferred into a container after pruning the roots and branches. The plants grown in pots may also planted directly in the container.

Containers

The containers should be unglazed shallow pots of various sizes and shapes. They may be round, rectangular or square, the size and depth depending on the plant to be grown. The containers should be simple and non eye catching.

Soil: The soil for potting should be fresh, well-drained, and on very rich in fertilizer. It should not be highly acidic or alkaline. Clayloam or loam of different structures and clump sizes, obtained by sieving the soil is used in potting. It should not be very sticky or sandy. Well-rotted leaf mould is mixed with the soil.

Soil Preparation: The soil is dried in the sun and sieved through meshes of at least three sizes. Large, medium and fine soil obtained by sieving are kept in separate containers. At the time of planting, the larger particles are placed at the bottom of the pot and a thin layer of medium-sized particles is spread over it. The tree is planted and the top soil consisting of small particles holds the plant in position and comes in close contact with roots.

Removal of Tree: The soil of the plant to be transplanted should be kept on the dryside to facilitate removal of the plant from the pot. It may be necessary to remove thick roots and also the tips of some fibrous roots with a pair of sharp shears. After pruning the roots, the shoots are pruned depending on the growth habit and the desired appearance of the plant in miniature form, when the branches are crowded in any part of the plants. In the case of unbranched seedlings growing terminally, the main stem is cut to a height, which may form a well-shaped Bonsai.

If the branches are small and well arranged, pruning is not needed but the terminal growth is removed to minimise the height of the plant and to encourage the growth of auxillary branches. In case the branches have already grown to a size too tall for a Bonsai, they are pruned to 1/2 to 1/3 the length depending on the rate of growth of the plant and its response to pruning. Generally broad-leaved plants and harder pruning than small-leaved conifers.

Planting in Container: The holes of the container are first covered with crocks or plastic net. Then a layer of large soil particles is placed with a thin layer of medium-sized soil above it. The tree is then placed in the soil. Before it is finally planted, the side presenting the best view should have been determined. The medium-sized soil is placed around the root region a consolidated by means of a bamboo stick without pressing too hard. Then a layer of fine soil is spread on the top, levelled and pressed gently. While planting, care should be taken to keep the base of the trunk from where the main roots rise above the soil and to place the plant in the middle of the container. After planting, watering should be done with the help of a fine rose from the top of the plant.

Care of the Plant: The container is kept on a platform in a cool and shady place for about two weeks to allow the roots to develop and the plant to establish. It is then gradually exposed to sunlight for longer durations beginning with the morning sun for two hours. The soil should never be allowed to dry up completely. In summer months with high temperature and low humidity, the plants may be placed in shade in the midday and afternoon. They also need protection from frost. The miniature plants should be arranged in a planned fashion on a plant form or on stands of different sizes and heights so that the collection of plants looks like a garden. Proper spacing between the rows will facilitate watering the maintenance of the plants.

Watering: Watering is very important for Bonsai, as they are grown in shallow containers with small amount of soil. They should not be allowed to dry as temporary wilting of the plants adversely affects their growth. If the roots shrivel due to shortage of moisture in the soil, the plants are likely to die. Excessive watering often causes poor growth and water-logging for a considerable period may cause rotting of the roots.

Pruning and Pinching: Removal of buds from Bonsai is an important practice. Pruning maintains the shape of the plant, stimulates branching and helps in the utilization of energy for the growth of other parts of the plant. The frequency and mode of pruning will depend on the growth habit rate of shoot growth, response to pinching and subsequent growth of auxiliary shoots.

In order that the plant may maintain an attractive appearance, pinching is not done at the same length in all the direction. In the case of two axillary branches growing in two directions, one may be pinched at the apical bud, while the other one is pruned up to several nodes below the terminal bud. If the plant tends to grow fast in a particular season or continues to grow in length throughout the year if may be necessary to pinch more than once a year. But frequent pruning leads to the formation of thin and weak shoots, which may wither in an unfavourable environment. Pinching or pruning should be done clean without damaging the shoot at the cut end It should be ensured that the plants remain in a fixed position and the soil and root are not disturbed at the time of pruning. Bonsai varies in shape and size. Following are common types:

1. A single tree with straight trunk
2. A tree with twisted trunk
3. A tree with slanted trunk
4. A tree with a large hanging branch
5. A tree with two trunks
6. A tree with several trunks
7. Several trees grown in a single pot
8. Tree grown on rocks

Arrangement of Branches

The arrangement of trunk and branches is also an important technique to be followed to develop an attractive Bonsai. It is done to improve the shape and to help in the manifestation of natural appearance of the plants. In order to maintain the space between two branches the lower one is suspended by a strong twine or wine.

Before using wire, the portion of the branch is wrapped with raffia where the wire is to be tied. To improve the arrangement of the branches on the plant, wire is to be used very carefully and the technique needs experience, skill and proper tools. Wire cutter and pincers are normally used for the purpose. Copper wire is better than iron wire, as it is softer and can be seen easily and does not rust.

Another very effective method of modifying the shape of the tree or arranging the branches is to tie wire around the trunk or branches so that they may grow in the desired direction and form.

Repotting: Bonsai needs repotting when the soil is completely exhausted or the container is filled with roots. In general, plants in the growing stage are repotted once a year, and a full-grown tree once every two or three years. Containers should be used after thorough washing and drying. While repotting, the old soil should be removed as far as possible and dead roots and ends of fine roots pruned. Unnecessary branches are also removed to improve the appearance of the tree. Then the Bonsai is planted firmly in the container using the similar type of soil mentioned earlier.

Application of Manure and Fertilizer

Fertilizer is necessary for Bonsai as the plant thrives and grows in a small container. One kilogram of mustard cake is diluted with six litres of water and allowed to decompose thoroughly. After about 3-4 weeks, the water above the decomposed manure is again diluted with 5-10 times the amount of water and the dilute solution is applied to the soil leaving the base of the trunk. Thoroughly powdered cake is lightly mixed with the topsoil slightly away from the trunk, in two or three areas using 1-2 tablespoonful of cake in each case. The plants should be manured in the spring and again in the rainy season when they show vegetative growth. During the growing season the plants can be manured once a month followed by watering.

Control of Diseases and Pests

Diseases and pests are common with Bonsai. Root rot is a common and serious disease caused by excess watering, drying of soil, poor drainage, direct contact with undecomposed organic matter, etc. the affected plant will show poor growth and decay of shoots. In the case of fungus infection either on shoots or on root, the diseased part should be removed at the earliest opportunity. Spraying of insecticides and fungicides should be a routine practice.

Lawns

Lawn is an essential and basic architectural feature of the modern garden. It has been observed that a well-planned lawn with its relationship to other furnishing of surrounding planting such as shrubberies, flower beds and specimen trees contribute greatly to the intrinsic value of the garden. It is a source of pride, everlasting pleasure, distinct charm, restful effect, and provides a great relaxation to the body and mind particularly after the hard labour of the day. Lawn also provides meeting place for social parties, marriage ceremonies, festival celebration and other gathering. As a matter of fact, no garden is complete and beautiful without having a good lawn, because the beauty of a garden largely depends on the condition of the lawn. Lawn is an inevitable 'must' because it serves as an important aid to beauty.

Lawns are always related to the design of the garden confirmating of mansion and houses in order to have perfect harmony with other garden features. For the excellent and brilliant growth of the turf, proper and thorough preparation of the soil is very essential. The subsoil should be retentive of moisture but must provide adequate drainage. The surface soil should be fertile loam at least 20 cm thick. Where the soil is very poor or gravelly the turf is liable to become patchy in dry weather, hence the topsoil of good texture should be of at least 20 cm thickness.

Selection of Site: The selected site for a lawn should get full sun and the best situation is the southern side of the house and the sites in southeast or southwest directions will serve the purpose. Grasses do not grow under the shade of tree.

Soil: Doob grass (*Cynodon dactylon*) can be grown on any type of soil. But for the establishment of a good lawn, fertile loamy soil rich in humus with good moisture holding capacity is most suitable. A slightly acid soil having pH between 5.5 and 6.0 is suitable for good growth of grass. Where the pH of the soil is lower than 5.5, 300 gm of chalk or ground limestone per sq. metre on sandy or a similar amount of slaked lime (calcium hydrate) on clay loam should be applied.

Preparation of the Ground: Thorough preparation of the ground is very essential, as the ground is the foundation on which lawn is established. A badly prepared lawn cannot be easily rectified afterwards.

In the hot weather, before the monsoon sets in, *i.e.* in April/May the ground should be dug up to a depth of 30 to 45 cm. The soil thus dug out and left in clods is subjected to the scorching rays of the sun for killing the weeds, insects, etc., and sterilising the soil. After few days when clods are partially dried, they are broken with the help of a bamboo or wooden rod. And thus the soil should be spaded 3 to 4 times at weekly intervals and each time the remaining clods are to be broken till the entire soil is brought to a fine tilth. The weeds including the roots and other foreign materials like rubbish, stones, etc. should thoroughly be picked up and removed.

Manuring and Consolidation of the Surface: If the soil is originally fertile it is better not to apply bulky organic manure into the lawn, because cow dung and farmyard manure contain weed seeds and its becomes very difficult to make the lawn weed free.

But where the soil is infertile, well-decomposed cow dung or farmyard manure should be applied to the ground at 100 kg per 100sq metre and thoroughly mixed with the top soil up to a depth of 10-15 cm. Once the soil has been disturbed it takes enough time to settle (consolidate) again. To ensure consolidation of soil the ground should be heavily watered 2 to 3 times at intervals. During this period weed seeds will germinate and nuts of *Cyperus* (motha) will sprout their sharp sword-like leaves. The weeds are removed before the next watering.

Levelling and Drainage of the Ground: Proper levelling of the ground is essential for a good lawn. After settling of soil, the surface should be levelled by filling depression and by repeated rolling. A gradual slope from the centre to the edges is always preferred as it facilitates drainage of excess water during the heavy rains and irrigation. For perfect levelling spirit level can be used with advantage.

Good drainage is also essential for maintaining a lawn in good condition. In small lawns where the soil is not so sticky, special arrangement of drainage may not be necessary. But for the bigger lawns, it is always advisable to have surface drains at the ends or sides of he lawns with a proper outlet for the excess water.

Selection and Planting of Grass: In most part of India, doob grass (*Cynodon dactylon*) is successfully grown in lawn. The planting material, either seed or grass should be free from weeds.

Lawn can be made at any time of the year provided irrigation facilities exist. In the tropical parts of the country it is better to sow the seeds after one or two showers in the beginning of the monsoon or after the close of the rainy season. But the grass runner should be planted in the beginning of the monsoon to get quick growth. Grasses are usually planted by dibbling, turfing or pasting of grass is also done.

(1) *From Seeds:* To ensure even sowing it is better to divide the entire ground by string into equal plots of 200 to 300 sq. metre. Prior to sowing, the rolled surface is scarified with a rake to a depth of 2.5 cm. The seed lot is divided for each small plots, *i.e.* at 500 gm/200 sq. metre and mixed with double their bulk of fine sifted soil and broadcasted carefully by hand. Then the soil is raked over into two directions so as to mix the seeds with the soil. If the ground is dry, it should be thoroughly rolled with a light roller. Then water gently with a water can having a fine rose to keep the soil moist. The seeds will germinate within 3 to 5 weeks.

    After few weeks when the grass will be 4 cm high, the first cutting should be done with a pair of sharp garden shears and the surface is rolled so that when the mower is used, it will go over a more or less even and low growth. Thereafter, no hand cutting will be required, as rolling and mowing at regular intervals will gradually improve the condition of the lawn.

(2) *Dibbling:* After thorough preparation of the ground small selected grass roots, free from weeds, obtained from either nursery or lawn scraping should be dibbled 7-9 cm apart preferably when the soil is slightly moist. Then the dibbled area should be lightly rolled and thoroughly watered so as to keep the soil moist till the grasses are established. After few week the grasses will be ready for cutting and in course of 4 to 5 months it will be a fairly compact lawn by frequent rolling, mowing and watering.

(3) *Turfing:* Turfing is the quickest method of forming a lawn. Turf or small pieces of grass should be cut uniformly thick from a place where the doob grass is free from weeds. They should be spread on the prepared ground side-by-side closely and slightly beaten down flat into position. Any cavity or interspaces should be filled with fine soil and entire turfed area then thoroughly rolled and watered liberally till the turf is established.

Another easy and successful method is to cut the off-shoots into small pieces and scatter them evenly over the prepared ground followed by a top dressing of fine soil and sand about 1 cm thick. Then light rolling is advisable if the ground is dry, followed by watering to keep the soil moist till the grass develops roots and new shoots come out. This method should be specially adopted in the rainy season and one is sure to have a brilliant and compact lawn during the next winter.

Once the lawn is properly established, it needs maintenance. To retain the verdure of the turf and its uniform colour, the lawn is raked, aerated, and top dressed with fertiliser and kept free from weeds.

Rolling and Mowing: The purpose of rolling is to bring the grass in contact with the soil and also to keep the ground level. After the first shower of rain, roll thoroughly with a medium roller both ways to make the level correct. Avoid rolling when the soil is too wet. Normally after every weeding or at weekly interval the lawn should be rolled to get an even surface. Regular mowing of lawn is also as essential as watering. Never allow the grass to grow more than 6 cm high. During the winter season most of the lawns require mowing once a week and some time twice when the growth is rapid after the application of fertiliser.

Watering: Watering must be thorough and as frequent as the weather condition requires. Frequent watering rather than heavy watering at long interval is beneficial for lawn. Sprinkler irrigation is the best method for watering lawn as it distributes water evenly over a good area and also saves much of the Mali's time. Fresh water should be used and brackish or saline water burns the grass especially during the summer.

In winter, dew is a great help to growing grasses and it is always advisable to take the full benefit by brushing the dew into the lawn with a light bamboo stick or a piece of rubber hose every morning.

Weeding: Both in old and new lawns weeds are found to grow along with the doob grass and if left uncared, in no time weeds will cover the entire lawn supressing the growth of doob grass and spoil the beauty. So just after the establishment of lawn one should be very careful about the removal of weeds at regular intervals or when required. Take out all plants that are growing on the lawn except doob grass. More frequent weeding is necessary during the rainy season than in the winter months. While weeding, care should be taken to remove the weeds including the roots, and never allow the weeds to produce seed.

Scraping and Raking: Due to constant rolling and mowing hard crust forms on the surface and the lower part of the grass becomes matted and woody. Therefore, scraping and raking are always beneficial and hence considered essential for the maintenance of lawn. In such cases the entire lawn should be scraped with Khurpi followed by May or early June. But in other cases where the condition of the lawn is good, only through raking in both ways will be sufficient to take out the old runners and loosen the soil. Then mow the grass throughly until the old stump of the doob is well-trimmed and the surface of the lawn cleaned and exposed to the sun. A top dressing of garden soil, coarse sand and leaf-mould in the proportion of 1:2:1 is beneficial. Hundred kgs of the mixture is sufficient to cover an area of 100 sq. m to a depth of 2 cm. Bonemeal @ 1 kg per 10 sq. m should also be applied.

Top Dressing and Manuring: Again in the month of August/September before the beginning of the cold season, a top dressing of garden soil, coarse sand and screened leaf-mould with the addition of bonemeal in the proportion mentioned above should be applied in the lawn followed by rolling. Ammonium sulphate at 1 kg per 50 sq. m should be applied once every month from October/November to February/March or a fertilizer mixture containing nitrogen and phosphorus may be used.

Trees

Tree is a woody plant with a spreading crown, whose single trunk exceeds diameter of 15cm and attains a certain height. To exclude shrubs, minimum height a height of more than 4 meters and up to 7 metres may be called a small tree. All trees are capable of producing seeds under favourable environmental condition and grow vigorously for many years. A tree may show the height and shape of a shrub in a climatic condition different from the natural habitat and temperature, light, humidity and moisture and nutritional status of the soil are found to play important role on growth and flowering of plants. By manipulating the environment, delicate plants can be successfully grown if one knows the requirements of those plants. Trees are very fascinating because of their graceful appearance and the abundance of bloom.

They are grown for their economic importance or aesthetic value or both. Fruit trees are planted for fruits and forest plantation for other economic products like timber, fuel, tannins, oils, gums, resins, waxes, spices, beverages, narcotics and drugs. The cultivation of trees for their aesthetic or recreational value is known as arboricultural. Here the individual tree is important in contrast to the wood as a crop as in forestry. They also exercise beneficial influences on climate and rainfall, regulate the water flow and prevent soil erosion. A large number of trees in our country are resplendent in riotous colours at the flowering time and are capable of transforming the landscape. The trees are the most permanent elements in landscape and a thorough knowledge of their ornamental properties, rate and mode of growth, their behaviour in different soil, situation and climate are essential. They should be planted carefully and thoughtfully for the benefit of height, shade colour and vertical emphasis.

We have got large number of indigenous and exotic flowering trees, which can be successfully utilized to beautify our cities, towns and villages. Along with the road plan, a plantation plan should be made and strictly adhered to. For the existing roads the dead and decaying trees should be replaced systematically according to a plan. Beautifully planted avenues with flowering trees are pretty with the colour and beauty. The trees should not be patchy due to lack of aesthetic sense of those maintaining the roads.

The value of trees as both labour saving and attractive to the residents of any private garden has become increasingly evident. Many trees burst into bloom beautifully, while others afford a pleasing contrast with their decorative foliage.

Although we have an abundance of flowering trees, selection of trees for private gardens which should create rhythm, accent, as well as balance in the garden and the dwelling place is rather difficult. One or two small flowering trees are often adequate selections and planting of trees should deserve just as much attention as is commonly given to the colour of the building, pool and paved path. The fruit trees should be planted to the back portion of the house where they are not visible from the entrance. A group of Plumeria or Cassia at the boundaries add charm and grace to the house. List of our indigenous ornamental trees will be very long and it includes many colourful flowering trees like Butea monosperma, Bauhinia purpurea, Cassia fistula, Lagerstroemia speciosa, Cochlospermum gossypium, Bombax malabaricum, Millingtonia hortensis, Dillenia indica and Saraca indica. Trees with ornamental foliage indigenous to India include Polyalthia longifolia, Putranjiva roxburghii, Melia azadirach, Mimusops elegi and Azadirachta indica.

Planting: The best time for planting trees is during the rainy season (June-September). In northern India planting can be done in January-February before the new growth starts and 1-2 years old saplings have better chance of survival and early flowering. Large pits (1 meter in depth and diameter) should be prepared 2-3 months before planting. The soil is mixed with adequate amount of organic manure and bone meal and allowed to settle by exposing to rains or watering the pits. The tree saplings should have straight stem and undisturbed main shoot. Before planting the ball of earth round the roots is cracked without damaging the roots and planted with root collar just below the ground level. During planting all dead and broken branches and roots are to be cut and removed.

The soil around the plant should be firmly consolidated after planting, watered thoroughly and also staked. In the open and in public places tree seedlings are to be protected in gabions till they grow fairly large. Distance of planting trees usually varies from 5-14 metres depending on the size of the plant. For better growth of the plant, the pits should be weeded and hoed to keep the soil loose and free from weeds. Most of the tropical trees have short dormancy period and monsoon is the season for maximum growth. In order to maintain the shape of the tree it is often necessary to remove the old branches. Crossed branches in the centre of the crown should also be cut off. On the stem or large branches holes are made by insect, rats, squirrel and birds or cracking of wood. The exposed surface of the hole is thoroughly cleaned by removing dead, and rotten wood, coasted with coated or fungicides and filled with cement.

Bauhinia Purpurea (Mountain ebony, Geranium tree)

Family: Leguminosae.

A nearly evergreen tree attaining a height 10-12m growing sparingly throughout India. Flowers large about 6cm across, in various tones of rose, purple, in few flowered clusters at the ends of the branches. It is a very hardy tree, flowers profusely from October to March, when leaves fall off but not completely.

Bignonia Megapotamica

Family: Bignoniaceae.

An evergreen tree of medium sizes growing up to a height of 10m and produces clusters of light mauve flowers. It is a quick growing tree in warm humid climate. The plant remains in bloom almost throughout the year but larger number of flowers develops from March to May.

Callistemon Lanceolatus (Bottle Brush)

Family: Myrtaceae.

The plant is also known as "Bottle Brush" because the flower-bearing portions of the branches resemble bottle brushes in shape. Flowers in densely crowded cylindrical spikes 5 to 10cm long, with long scarlet stamens projecting stiffy outwards. The flowering may be observed more or less throughout the year, especially from February to November.

The trees have pendulous branches, often grown on roadside and in gardens.

Azadirachta Indica (Indian liac, Margosa tree)

Family: Meliaceae.

A medium sized almost evergreen tree usually maintaining a height between 10-15m. The tree is very popular in India because of its high medicinal properties. The dark evergreen foliage gives a very showy appearance and it is also grown as an ornamental tree.

Erythrina Indica (Coral tree)

It is a tall deciduous tree reaching a height up to 18m; bark is smooth, yellowish or greenish grey. Flowers large, red 5 cm long, pea shaped in dense recemes 16-20cm long that appear in February-May. The tree is commonly planted in villages and along roadsides as shade trees and is used to make close hedges because it grows readily from massive cuttings and the prickles ward off intruders.

Delonix regia (Gulmohar, Flame tree, Peacock flower)

Family: Leguminosae.

It is a large deciduous tree, native of Madagascar and reaching a height of 12-20 m with spreading branches, umbrella shaped crown and greyish bark. Flowers 5cm wide, scarlet in colour with a mild scent, borne on long stalks in short axillary recemes forming panicles on the new shoots; other portion or branches are bare of leaves. It is one of the most beautiful and common flowering trees grown in India and very suitable for parks, roadside and also large private gardens.

Ficus Religiosa (Peepal)

A huge tree with greyish bark. Leaves smooth shining, broadly ovate, apex long and narrow, 10-18cm long. The tree is indigenous in Bengal and Burma and is cultivated all over India. Handsome dense foliage on the spreading branches gives a cool and pleasant shade under the tree.

Ficus bengalensis (Banyan)

Family: Moraceae.

A large evergreen tree, may attain a height of 30m, branches spreading, almost horizontal. The tree is indigenous near the foot of the Himalayas and Western India and is commonly planted all over the country.

Orchids

Orchids are considered as the most beautiful flowering plants for the exquisite beauty of the flowers, variety of fragrance, brilliance in colour, unusual shapes, variation in form and attractive growth habits. There are about 24,000 species and 32,000 hybrids of orchid.

Development of new hybrids and commercial production of cut flowers in orchids are expanding rapidly in the U.S.A. Europe, Thailand, Malaysia and Singapore. There is immense scope for improving orchids in India, because large number of species are native to this country and many of them have already proved to be important parent plants and contributed in the production of several outstanding hybrids in the world. Due to the diversity of environmental condition in India, it is possible to grow all types of orchids in suitable places without the control of environment.

Forest is the natural habitat of orchids. More or less similar environment can be created by growing the plants in greenhouse and protecting them from direct scorching sun, dry wind and by maintaining high humidity.

Warm Climate Orchids: Orchids which suit warm climate condition and can be successfully grown in ordinary greenhouse include the numerous hybrids of Cattleya, Dendrobium, Onicidium, Phalaenopsis, Rhynchostylis and Vanda, Orchid species producing beautiful flowers *e.g.* Aerides longicornu, A. multiflorum, Arachnis crispum, A. maculosum, A. odoratum, Anoectochilus roxburghii, Arachnanthe catchcartii, Arundinia bambusaefolia, Calanthe densiflora, C. masuca, Coelogyne flaccida, C. ochracea, Dendrobium aggregatum, D. densiflorium, D. devonianum, Phaius macutatus, P. wllichii, Phalaenonsis manii, P. parishii, Pleione maculata, Renanthers imschootiana, Rhynchostylis retusa, Saccolobium ampullaceum S. dasypagon, S. trichromum, Thunia alba, Vanda coerulea, V. coerulescens, V. cristata, V. teres and many others grow and flower well in the above enviornment.

Orchid House and its Management: A free standing flat-roof orchid house shaded by spit bamboo or wooden batten is recommended for housing of orchids suitable for warm climate. The temperature range suitable for most of those orchids is 65 to 85^{0}F. For satisfactory growth of orchids, atmospheric humidity should not be less than 30 per cent at night and 70 to 80 per cent during the daytime. Monopodial orchids like Vanda, Phalaenopsis require high humidity, whereas sympodial type *e.g.* Cattleya, Laelia or those with leathery leaves need less humidity. The atmospheric humidity will increase if small tanks or lily pools are located inside the orchid house and the floor space is covered with sand, soil, cinder, etc. instead of concrete. Free circulation of air is needed for the orchids to grow and flower and light intensity ranging between 1500 to 2000 feet candle in midday is good enough for most of the orchids.

Seed Sowing and Care of Seedlings: Seedpod of orchid grow after fertilization, and ripens in six months to one year. After ripening the seeds are collected and stored in a cool and dry place or in a desiccator. Millions of powdery seeds are released from each pod and they contain little or no food to nourish the embryo. Under natural condition, the seeds germinate when they find a right pocket of decaying vegetable matter on the trees. Seeds of orchids are germinated and seedlings grown in culture media containing agar, inorganic nutrients and sugar. Disinfected seeds are sown in sterilized flasks containing agar-nutrient media and the seedlings grow for 8 to 12 months before they are transferred.

The seedlings are removed from the flask and planted in the community pots, 7 to10 cm in diameter which hold about 20 to 25 small plants. The compost used for seedlings in the community pot is a mixture of equal parts of the finely chopped tree fern and dust-free crushed bark or moss. A shady but well aerated location in the greenhouse will promote the growth of seedlings. In the community pot, the seedlings are watered daily and during the summer months they may be sprayed with water two to three times a day.

With the increase in size and vigour, each plant is transferred in a small pot, using the same compost recommended for larger plants, but at this stage they benefit by feeding with the weaker concentration of fertilizer solution.

Vegetative Propagation: Besides multiplication by seeds, commercial method of vegetative propagation of hybrids of Cymbidium, Phalaenopsis and Cattleya is done by meristern culture and large number of plantlets develop from a small piece of growing apex. Amateurs, however, propagate their plants by offsets, air layering, cuttings and division. Offsets develop from Dendrobium and some Epdiendrum, which can be detached and planted in small pots.

Air layering is practised on the monopodial types like Vanda. A slant cut is given halfway in the stem and wrapped with sphagnum moss. When roots are noticed in the moss, the upper portion of the plant with roots is detached and potted. In Arachnis, Renanthera and to a lesser degree in Vanda propagation is done by cutting. As these large-sized plants produce adventitious roots, the stem is cut in section 3 to 4 nodes, placed in a cool and dry place for healing of wound and allowed to root in moist sand or damp sphagnum moss. Plants of many genera such as Cattleya, dendrobium, etc., produce new growth from a lead and they can be propagated by division at the time of repotting.

Potting and Compost: A vigorous and healthy root system often indicates good vegetative growth of the plants, which largely depends on the pot compost. Ideal rooting media will provide high degree of porosity and ensure adequate oxygen for root respiration. Water should drain out freely through the media and it should be resistant to rapid decomposition and decay.

Depending on the growth habit, *i.e.* terrestrial or epiphytic, orchids are potted in a wide variety of media and compost. Epiphytes like Cattleya, Epidendrum, Phalaenopsis Vanda, Dendrobium, Rhynchostylis, etc., are planted on a very light rooting media, consisting of various kinds of tree fern fibre or on larger pieces of hard charcoal. Terrestrial orchids like Phaius, Calanthe, Thunia, etc., thrive best in the mixture of leaf-mould, loamy soil, sliversand, dried cowdung manure, charcoal and chopped tree fern fibre. Epiphytic orchids are best grown in speciality designed orchid pots with holes at the bottom and slits or perforations on the sides. Monopodial epiphytes like Aerides, Phalaenopsis, Vanda, etc., are particularly suited to basket culture because of the large aerial roots produced from these plants and straight growing habit. Many orchids are conveniently grown on the branches of the trees, charred wooden slabs or tree fern blocks.

Orchids grow better if undisturbed, Phalenopsis, Vanda, Laelia and some species of Dendrobium should not be repotted unless it is absolutely necessary. Cattleya and many Dendrobium species and hybrid need repotting when roots become pot-bound.

Watering and Spraying: As orchids are grown in a light and porous compost, watering is very important. Atmospheric humidity influences evaporation of moisture from the compost and orchids, in general, prefer high relative humidity. In a dry and will-ventilated atmosphere, damping down of the floor of the greenhouse, frequent overhead sprinkling of water will increase the humidity. But in high humid atmosphere watering should be less frequent. Alkaline water is injurious to orchids and slight acidic water or at pH up to 7 should be used.

Newly potted plants should not be watered very frequently but the compost is kept moist by fine spraying. As new root emerges and growth starts, the plants need more frequent watering and it should be decreased again after flowering. Fine spraying of water is also very beneficial to plants during the period of growth or on warm dry days.

Nutrients: It has been observed that growth and flowering or orchids are improved markedly by the application of fertilizers in liquid form and a number prepared fertilizers are available in the orchid-growing countries. A balanced feed on nitrogen, phosphate and potash in the ratio of 10:12:10 and very small amount of magnesium, calcium, manganese, iron, boron and zinc has been found very effective on a large number of species and hybrids.

For mature and flowering plants. Two table spoonful of the above fertilizers mixed in 10 litres of water is sprayed once a week, while a more dilute solution is used on seedlings. Leaves and rooting media should be thoroughly sprayed with the fertilizer solution.

Diseases and Pests: Orchids are less subjected to the attack of pests and disease. Scale insects, mealy bugs, green fly, thrips, red spider and snails may cause considerable damage, fly, thrips, red spider and snails may cause considerable damage, if they are not controlled in time. Application of Rogor or Malathion is very effective to keep the orchids free from pests.

Die-back is a serious disease in orchids which starts in rhizome and if left unattended, it spreads to other plants in the orchid house. Orthocide 50 and Cossan are recommended for controlling fungus diseases on orchids. Virus infection is also common in several species and varieties of orchids. Sometimes black spots appear on leaves and flowers turn yellow and drop off. This is not caused by fungus but due to faulty culture like over-watering insufficient ventilation, too much of light or very dry atmosphere.

Cleanliness of the greenhouse and regular attention to the plants are very important to keep the plants free from diseases and pests.

Landscape Gardening

Landscape gardening is an aesthetic branch of Horticulture which deals with planting of ornamental plants in such a way that it creates a picturesque effect. It is a very fascinating and interesting subject. There are several definitions and expressions to define this subject.

According to Chambers' dictionary, the definition of landscape is the appearance of that portion of land which the eye can view at once and landscape gardening is the art of laying grounds so as to produce the effect of a picturesque landscape. Landscape gardening can be defined as the decoration of a tract of land with plants and other garden materials so as to produce a picturesque and naturalistic effect in a limited space. So landscape may or may not include plants. According to Bailey, Landscape gardening is the application of garden forms, methods and materials to the improvements of the landscape and the landscape in this connection is any area large or small on which it is desirable to develop a view or design.

Landscape gardening can also be defined as the beautification of a tract of land having a house or other object of interest on it. It is done with a view to create a natural scene by the planting of lawn, trees and shrubs. Landscape gardening is both an art and science of the establishment of a ground in such a way that it gives an effect of a natural landscape. It can be also defined as the imitation of nature in the garden. It can also be defined as improving of total living environment for the people. The expression of landscape may be gay, bold, retired, quiet, etc. This expression will conform to the place and the purpose. It should be a picture and not a collection of interesting objects.

Since the landscape gardening is the making pictures on the ground with plant and other material, landscape designer should be proficient in art, ornamental gardening, ecology and physiology. He should be an architect and engineer to appreciate the relationship between plant form, colours and buildings.

Natural Elements of Landscape

Different types of landscape depending upon prevailing geographical and agroclimatic conditions characterize Earth. There are mountains, hills, glens, valleys, seas, rivers, forests, plains, deserts, lakes, swamps, streams, etc. which comprise major part of natural landscape. At certain points, there is harmony between natural elements like ground forms, vegetation and even animal life. The landscape of such place is beautiful and conveys the feeling or mood of the landscape character like exhilaration, sadness, ceriness or awe. There are many qualities of natural landscape beauty like the picturesque; the ethereal, the serene, the delicate, the idyllic, the graceful, the majestic, the bold etc.

Man has copied the natural elements for improving landscape around him and converted certain areas in the form of garden for his pleasure.

Landscaping Principles

General principles of landscaping are as follows:

i. The ideal landscape garden is like ideal landscape painting which expresses some single thought or feelings. Its expression may be gay, bold, retired, quiet, etc.

ii. Beauty and utility should be harmoniously combined.

iii. Area should be divided into different parts and plan should be conceived for each area. Overall plan should be such that the observer catches the entire effect and purpose of the plan without stopping to analyse its parts.

iv. Simplicity of design should be aimed at in the execution of the plan.

v. An ideal landscape should have open space.

vi. Let the garden and building merge into each other. There should not be stopping abruptly particularly in front of building. The view of garden from the windows and doors should be very attractive. Planting around the building, climbers against wall and on the porch, decoration of verandah and rooms with attractive foliage, flowering plants, hanging baskets serve to unique the building with garden. Every part of the compound should be planned in such a way that it gives surprising effect to visitor.

vii. Over crowding of plants and objects should be avoided.

Factors Affecting the Landscape Design

There are several factors which affect the making of suitable design for particular site. These factors are:

- *Human Choice:* Man's ultimate desire is to make his living pleasurable and surroundings confortable. His dominance in making designs and selection of plant material is very well evident. Therefore, different styles of gardening have come into existence.
- *Site:* This is an important factor and according to site, suitable design is made. In formal style gardening, the site is selected according to plan. Topography of the site also affects the design.
- *Views:* Distant views of mountains, hills, woods, valley, etc. are preferred from the place of garden.
- *Heritage:* One inherits the knowledge of botany and aesthetic sense and uses accordingly. Our rich heritage teaches us to use flowers and fragrant trees to improve the surroundings.
- *Climate:* The climate of particular place affects the selection of plant material accordingly. Ideally suited plant material according to climate should be selected.
- *Soil:* According to characteristics of soil types, suitable plants should be selected.

Garden Styles

Garden styles have been changed from time to time with the new ideas and necessities. Broadly, the styles of gardening are grouped into three categories *i.e.* (i) Formal style, (ii) Informal style and (iii) Free style of gardening

(i) *Formal style:* Main features of this style of gardening are: First plan is made on the paper and then land is selected accordingly. Plan is symmetrical. These types of gardens are of geometric design *i.e.* squarish or rectangular. Therefore, the roads are cut at right angle. It has some sort of enclosure. Flower beds are also of geometric shapes. The arrangements of tree and shrubs are necessarily geometrical and kept in shape by trimming and training. Other features like fountains, water pools, cascades, etc. are used for further attraction. The examples of such style of gardening are Persian gardens and Mogul gardens.

(ii) *Informal Style:* This style reflects naturalistic effect of total view and represents natural beauty. This style is just contrast of above formal style. In this, plan is asymmetrical and according to the land available for making garden. Roads, paths are made curvaceous and bending. Water bodies are made of irregular shapes. Hillocks are made to create natural mountainous scenery. Flower beds are made of irregular shapes suiting to surroundings. Plants are allowed to grow in natural form and instead of trimming, annual pruning is done. Japanese gardens are the best example of this style of gardening.

(iii) *Free Style of Gardening:* This style combines the good points of both formal and informal style of gardening. Rose garden of Ludhiana is an example of this style of gardening.

Art Principles of Landscape

Landscaping is making of pictures with plant material and, hence, its principles are same as those of art. They are as follows:

- *Rhythm:* Repetition of same object at equidistance is called rhythm. It can be created through the shapes, progression of sizes or a continuous line movement, rhythm creates movement to the eye. In gardens, generally trees of single species of equal height and shape are planted to create this effect. In Mogul gardens, fountains and water canals have also been extensively used to create such effect. Now-a-days other objects like lights are also used to create the effect of rhythm.
- *Balance:* It is very important to maintain the balance on both sides of the central line. The principle involved in making balance of see-saw game can help in understanding this. Equal weights can be balanced only when they are equidistant from the centre. If weights are unequal, the heavier must move towards centre for making balance. The balance may be formal, informal or symmetrical types. Imbalance will look lopsided and will distract the attention. In making the balance with the plants, their form, colour, texture etc. are kept in view.
- *Accent or emphasis:* The accent or emphasis is created in the gardens to avoid the monotonous view. It is the method to stress the most important thing. This also serves as the centre of attraction. Mostly unusual objects like tall fountain, tree, statue etc. are used to create the effect of accent or emphasis. In English gardens, statues have been used extensively to create such effects.
- *Contrast:* This principle is most useful in emphasising the best features of an object. It can be very easily understood by following contrast colour theory.

Against green background, a fleck of scarlet colour will make a contrast and will make scarlet colour prominent. In nature this is very common. Other contrast colour can also be used. Similarly, weeping growth habit against upright growth, dwarf against tall, rough texture against soft texture etc., are some of the examples which can be followed. It is also very important that one of the two contrasting objects must clearly dominate each other. In this way, one becomes feature whereas other acts as supporting background. The contrasting elements of equal power may create visual tensions.

- *Proportion:* It is the relation of one thing to another in magnitude. When two or more objects are put together the proportions are established. In a landscape design, space provided for lawn, paths, herbaceous borders, shrubbery border, trees, buildings and other garden objects should be in a right proportion. It will create harmonious effect and look better. Such effects can be noticed in Persian and Mogul gardens. Out of proportion allotment of area in garden will distract the attention. Proportion helps in space organization.
- *Harmony:* It is an overall effect of various features, styles, and colour schemes of the total scene. The degree of harmony or unity of various elements of landscape is a measure induced in us and is called as beauty. Therefore, the beauty can be defined as the evident relationship of all parts of a thing observed. When different parts of landscape are correctly placed in right way, produces a harmonious effect. Such landscapes create picturesque effect and appeals to visitors. On the contrary, the absence of harmony or lack of unity is ugliness.

Some Important Landscape Gardening Terms

Axis: It is an imaginary line, which divides garden into two parts. This also connects two or more points. It is presented in the form of a path, line of fountains or trees, etc. If this axis divides garden into two equal parts, it is called central axis. In formal style, axis is central whereas in informal style, it is oblique. This axis controls the movement in the garden from the entrance to the terminal. An axis in garden is directional, orderly or dominating.

Symmetrical Plan: In this plan, the different objects are in equilibrium about a central point or on either side of an axis. Symmetrical plan or formal plan is synonymous of beauty and is pleasant and handsome. This is because the symmetry is to be associated with plan clarity, rhythm, balance, unity, etc. Symmetrical plan being precise and disciplined, it requires precision in detail and maintenance and bold in concept.

Dynamic Symmetry: In such symmetrical plan, each pole generates its own magnetic field and between these two fields there is a field of dynamic tension.

Asymmetrical Plan: In such plans, there is absence of symmetry on both the sides of axis but balance, unity and harmony are maintained.

Circulation in Landscaping: In landscape gardening it means a pathway from entrance to terminal point. Circulation varies with the style of gardening and topography. The more of circulation patterns, it has more points of views and attraction. Common circulation patterns are depicted.

Vista: It is a three dimensional confined view of terminal building or dominant element of feature. It may be natural or man-made. Natural vistas are very common around the lofty mountains and snowy peaks. Overall effect of vista may be of its characteristics. It may be calm or induce motion. There are three different parts of a vista *viz.*

- It should be subject to a close control.
- It should have a viewing station to see object or objects.
- It should have intermediate ground. These three should be satisfactorily united and thus result into an effect of totality.

Chrysanthemum

Chrysanthemums are next to roses in popularity and have been in cultivation for more than 2,500 years. There are thousands of varieties now in cultivation in different countries and more than 3,000 varieties are being grown in English gardens alone.

Classification: Chrysanthemum varieties are classified into seven main groups, namely, the Incurved (like a perfect ball), Reflexed (with drooped florets), Incurving (in which the petals incurve loosely and irregularly), Anemone (having single petals and a tubular central disc), Pompon (with very small-sized flowers), Singles (having a central disc and five petals or ray florets), Miscellaneous like Spider (with a hook at the tip of petals), Spoon (with a spoon-like tip of petals), Koreans (having small single or double flowers with a visible central disc) and Rayonnantes (having quilled petals).

Varieties: There are numerous varieties of chrysanthemum, the flowers of which may be white, cream, yellow, red, terracotta, bronze, maroon, lilac, mauve, pink or purple. A few commonly grown varieties are Snowball (white, incurved), Sonar Bangla (cream, incurved), Rose Bowl (deep rose-pink, incurved), Alfred Simpson (crimson-red, bronze reverse incurved), Pink Cloud (pink, incurved), Ajina Purple (rose -purple, incurving), Bronze Turner (bronze, incurving), Coronation Pink (pale pink, reflexed), Duke of Kent (white, reflexed), Mahatma Gandhi (mauve, tubular, reflexed), Kasturba Gandhi (white, tubular, reflexed) and Valiant (large, white, reflexed).

A few other important varieties include tubular, Kikubiori (large, yellow, incurved), Angel Bell (large, soft, pink, incurved), Shen Meigetsu (yellow, incurved), Grape Bowl (wine-red, incurving), Tokyo (white, spider), Rupasi Bangla (white, spider, dwarf), Flirt (deep maroon, double Korean type), Birbal Sahani (white, globular, pompom), Mountaineer (large, yellow, incurved), and William Turner (white, incurved).

Cultivation: During early February, the plants are beheaded after the flowering is over by cutting the stem about 15-25 cm above the soil. After some time when new suckers appear at the base of the stool, these are separated and planted in small 10cm pots. Each sucker has its own root. The potting mixture consists of one part each of sand, soil and leaf mould and a trace of wood ashes

The second potting is done at the end of April and the suckers are transferred to a bigger pot of 15cm which is filled with a richer soil mixture containing one part sand, one part soil, two parts leaf-mould, a quarter part wood ashes and one tablespoonful of bonemeal.

The third and final potting is done in August when the plants are shifted to 25-30 cm pots. The potting mixture consists of one part sand, one part soil, two parts leaf-mould, two parts cowdung manure, a quarter part each of small pieces of wood charcoal and wood ashes and two tablespoonfuls of bonemeal. During May and June, the young plants must be protected from strong sunshine and from heavy rain during the rainy season.

Stopping and Disbudding: By late May or early June, the young plants are pinched (stopped) by removing the tip of the main stem at a time when the young lateral shoots or "breaks" are just appearing in the leaf axils. Stopping will induce lateral growths to develop from the leaf axils.

The number of main stems one would like to keep can now be decided upon. Generally one, three or six stems are retained for obtaining exhibition blooms. At the end of each stem there develops the first crown bud which is allowed to develop and the lateral growths arising in the leaf axils are disbudded or removed.

The date of blooming depends mainly upon the time of starting the suckers or cuttings and dates of stopping and disbudding. The plants will need staking during October.

Manures and Fertilisers: Ammonium sulphate, about 30-35g can be mixed in two gallons of water and a pint of the mixture applied to each plant during July-August. Soon after the appearance of flower buds, sulphate of potash may be applied in the same way as ammonium sulphate.

About one tablespoonful of superphosphate should be mixed in the soil at the time of final potting. Nitrogen promotes healthy vegetative growth, while potash helps in developing strong and stout stems and flowers and phosphorus produces a good root growth. Liquid manure is prepared by dissolving ½ kg of fresh cowdung in five gallons of water and after four or five days when it has fermented, it is diluted with water to a very light tea colour. Liquid manure can be applied once a week after the appearance of flower buds till the flowers are half open.

Watering: The young plants require frequent watering in summer but less during the rains. Over-watering should be avoided.

Pests and Diseases: The Chafer bettle grubs appear usually in July and August at the base of the pot and cause the wilting of plants. These grubs should be removed by hand and destroyed. It is also useful to mix a little 5 per cent BHC and DDT dust mixture in the soil. The other insect pest is the aphid which appears in the cold months and sucks the sap from the leaves. Spraying Malathion, 2ml in one litre of water, is effective in controlling aphids.

Among diseases, wilt and powdery mildew are important. The wilted plants should be uprooted and burnt as soon as they appear. To control powdery mildew, the white mould on leaves, dusting with sulphur is useful.

Vegetables

Most seeds normally remain viable for 2 or 3 years if stored under good conditions. To grow best vegetables either for home garden or for growing commercially try only certified seeds with, trueness to type and freedom from certain diseases. Good seeds are clean, viable, free from disease and true to the name variety. Therefore, buy only from seed firm of known integrity. High yielding, high price seeds should have 90% germination. Seeds of 50% germination is very poor.

Germination of Seeds: For germination of seeds adequate moisture, temperature and aeration are essential. The requirement of temperature for various vegetable seeds varies markedly. Some seeds do not germinate at low temperature while some others at high. Usually germination is optimum in between 40^0 F and 60^0 F. Seeds absorb moisture and swell and vital activities starts. Respiration begins and energy is supplied and this requires oxygen. Aeration is essential to supply oxygen. If supply of water is more or over wet aeration is poor and may hinder in germination.

Seed Treatments: Vegetable seeds are usually grown in nursery beds or in boxes. There are three types of seed treatments to control diseases.

The first eliminates organisms present on the surface of seeds. Calcium hypochlorite, mercuric chloride and bromide water may be used. The second disinfectant eliminates organisms present within the seeds.

For this hot water, formaldehyde and mercuric chloride are effective. In hot water treatment dry seeds are immersed in hot water 45 to 55^0 C for 10 to 15 minutes. The third treatment protectants are fungicides to protect seeds from soil fungi. Nursery soil should be drenched before sowing of seeds. Commercial materials are also now available to treat seeds like Captan, Cerason, Thiram, Bavistin, Agrosan GN and antibiotics such as Agrimycin (0.01%), Streptocyline (0.01%) etc.

To secure healthy seedling the seed treatments should be invariably done. If the outside nursery bed contain worms, maggot and such insects a combination of insecticides and fungicides should be applied like Aldrin, Dieldrin with Captan, Thiram etc. All chemicals should be handled carefully according to directions of the manufacturer.

If boxes are used for growing seedling the soil should be disinfected by heating the soil to 180 F for atleast one hour. If nursery beds are used the soil may be disinfected by 40% Formalin. One part formaldehyde (40%) mixed in 50 parts of water and saturates the soil. After 24 hours of treatments the seeds may be sown. If captan and formaldehyde is used, the nursery soil may be drenched about 10-12 days before seed sowing.

Preparation of Nursery Beds: The nursery beds should be prepared before the soil treatments. The nursery beds should be of one meter width and length four to five meter. The soil should be cultivated to a find tilth. All weeds, stones etc. should be removed. The beds should be raised to 15 to 20 cm. heights. If necessary, shade or shelter may be provided. Usually total bed area of 80-100 sq.m. is sufficient for most of transplanted seedlings to cover one hectare.

NPK nursery mixture may be applied along with compost on FYM during preparation of soil. After complete preparation and levelling the beds should he given soil treatment as mentioned above.

Sowing of Seeds: The vegetable seeds should be sown in nursery beds in lines at 1.5-2 cm. below surface soiled at a distance of 5-6 cm. The distance from line to line should be 10-15 cm. After sowing cover the seeds with sieved compost very lightly.

The beds should be watered with a sprinkler. The above distance from plant to plant or row to row may be changed according to size of seeds, kind of vegetables and type of seedlings. Similarly the depth of covering with compost varies with the kind of seed. Very fine seeds may be dusted over the nursery bed. Covering of other seeds may be one to two times their minimum diameter. During germination and after sprinkle water in the beds but no excessive. Over watering creates high humidity and poor aeration and contributes to "damping off" disease.

To transplant the vegetable seedlings to its permanent location the seedlings should be stocky, healthy and vigorous rather than spindly, weak and elongated plants. Moderate temperatures produce stocky growth. To harden some vegetable seedlings, seedlings may be transplanted one or two times to another nursery bed. In large scale condition watering may be withheld two to three days ahead of transplanting to permanent site.

The young seedlings in the nursery bed should protected against diseases and pests. Regular spray to young plants with Malathion 50 EC or Dithane M-45 or Sevin (0.2%) similar material at 10-15 days interval is necessary.

Most of the vegetable seedlings in the nursery bed is ready for transplanting when the seedlings are 4 to 6 weeks old. During uprooting every care should be taken not to damage roots. Light watering should be given at least 12 hours before lifting.

Bibliography

Arguello D. S.: *Production, Post-harvest Management and Exportation of Tropical Fruits in Costa Rica*, Wageningen, Technical Centre for Agricultural and Rural Cooperation, 1992.

Ashraf, Syed Mahboob: *Handbook of Fruit and Vegetable Products*, Agrobios, Delhi, 2008.

Barbeau G.: *Tropical Fruits in Nicaragua*, Managua, Nicaragua Ministerio de Desarrollo Agropecuario, Agraria, 1990.

Bhutani, R.C.: *Fruit and Vegetable Preservation*, Biotech Books, Delhi, 2003.

Bosworth, B.: *Capital Formation and Economic Policy of Vegetable Production.* Brookings Papers on Economic Activity, 1982

Chadha K. L. and Pareek O. P.: *Advances in Horticulture: Fruit Crops*, New Delhi, Malhotra Publishing House, 1993.

Collymore L.: *Fruit Production in Barbados*, Port of Spain, Trinidad and Tobago, 1996.

Coste R.: *Coffee: the Plant and the Product*, London, MacMillan, 1992.

Dabholkar, A.R.: *General Plant Breeding*, Concept, Delhi, 2006.

Dharamvir Hota: *Modern Biotechnology in Plant Breeding*, Gene Tech Books, Delhi, 2007.

Doijode S. D.: *Seed Germination in Fruits*, New Delhi, Malhotra Publishers, 1993.

Featherly H. I.: *Taxonomic Terminology of the Higher Plants*, USA, Iowa State College Press, 1954.

Gunjate R. T. and Tawde A. B.: *Propagation and Rootstocks of Mango*, New Delhi, Malhotra, 1993.

Gupta, S.K.: *Disease Problems in Vegetable Production*, Scientific, Delhi, 2006.

Herminie Broedel Kitchen: *Soils and Crops : Diagnostic Techniques*, Satish Serial Publishing, Allahabad, 2004.

James S Shoemaker: *Culture of Vegetable Growing*, Asiatic Pub, Delhi, 2006.

Jeffers P.: *Evaluation of Four Onion Varieties in Montserrat*, Plymouth, CARDI, 1992.

Julia F.: *Fruits of Warm Climates*, Miami, Julia F. Morton Publisher, 1987.

Kataria, T N: *Plant and Crop Physiology*, Pearl Books, Delhi, 2008.

Kumar, N.: *Breeding of Horticultural Crops : Principles and Practices*, New India Pub, Delhi, 2006.

Morris M. L. : *Maize Seed Industries in Developing Countries*, Colorado, Lynne Rienner, 1998.

Narang, R.K.: *Fruit and Vegetable Preservation Techniques*, APH Pub, Delhi, 2010.

Oldham P.: *Cost of Production of Major Tree Crops in Dominica*, Roseau, Ministry of Agriculture, 1991.

Rao , P. Venkateshwara: *Dairy Farm Business Management*, Biotech Books, Delhi, 2008.

Rehm Sigmund: *Multilingual Dictionary of Agronomic Plants*, Boston, Kluwer Academic Publishers, 1994.

Singh, Kirti: *Emerging Scenario in Vegetable Research and Development*, Research Periodicals and Book Publishing House, Delhi, 2001.

Singh, S K : *Biotechnology, Plant Propagation and Plant Breeding*, Campus Books, Delhi, 2008.

Somasundaram Rajarathnam: *Advances in Preservation and Processing Technologies of Fruits and Vegetables*, New India Publishing Agency, Delhi, 2011.

Srivastava, A.P.: *Mechanization of Vegetable Production and Post-Harvest Management*, Westville Pub, Delhi, 2009.

Thomas E.: *Fruit Production in St. Kitts and Nevis*, Port of Spain, IICA, 1996.

Tripathi, Pramila: *Botanical Pesticides in the Management of Post-harvest Fruit Diseases*, Daya, Delhi, 2005.

Verma, L.R. and Sharma, R.C.: *Diseases of Horticultural Crops: Fruits*, Indus, Delhi, 1999.

Walia, S. and B.S. Parmar: *Pesticides, Crop Protection and Environment*, Oxford & IBH, Delhi, 1995.

Yadav, D.S.: *Advances in Vegetable Production*, Researchco Book Centre, Delhi, 2005.

Index

A

B

C

N

O

P

R

S

T

V

W

Y

□□□